Modern Knitting Technology

Modern Knitting Technology

Sam K Zachariah
Sreekala K

2026
Agri Horti Press
New Delhi-110 084

ISBN: 978-93-49690-23-3 (Hardbound)

Published By

Agri Horti Press

New Delhi

E-mail: agrihortipress@gmail.com

Digitally Printed at : Replika Press Pvt. Ltd.

PREFACE

Knitting, one of the oldest forms of textile production, has evolved significantly over the years. From hand-knitted fabrics to highly automated, precision-driven industrial processes, modern knitting technology has revolutionized the textile industry, enabling the creation of intricate, high-performance fabrics used in fashion, medical applications, and technical textiles. The seamless integration of design, innovation, and technology has led to the rapid growth and diversification of the knitting sector.

This book, titled *Modern Knitting Technology*, is designed to provide an in-depth exploration of the principles, techniques, and technological advancements in the field of knitting. It is intended for students, engineers, textile professionals, and industry experts who wish to understand the modern applications and innovations in knitting technology.

The book starts by laying a solid foundation in the fundamentals of knitting, including the various knitting techniques, types of knitting machines, and the mechanics of yarn manipulation. It then moves on to explore the diverse knitting systems, from traditional weft knitting to advanced warp knitting technologies. Each type of knitting machine and process is covered in detail, with explanations of their functions, capabilities, and ideal applications.

In later chapters, the book explores the latest developments in knitting machinery, such as computerized knitting systems, 3D knitting technology, and circular knitting machines, along with the integration of automation and digitalization in the manufacturing process. A significant focus is placed on how these innovations improve the efficiency, quality, and flexibility of production while reducing material waste and energy consumption.

Further, the book examines the wide-ranging applications of knitted fabrics in various industries, from apparel and sportswear to medical textiles, automotive fabrics, and technical textiles. Case studies and examples of cutting-edge projects demonstrate how modern knitting technology is shaping the future of textile design, functionality, and performance.

With a blend of theory, technical insight, and practical guidance, *Modern Knitting Technology* offers readers a comprehensive understanding of both the art and science of knitting in the context of today's fast-evolving textile landscape. Supported by illustrations, tables, and real-world examples, this book serves as a valuable resource for those seeking to stay at the forefront of the knitting industry.

We hope this book will inspire further research, innovation, and a deeper appreciation of the role that modern knitting technology plays in the world of textiles.

Sam K Zachariah

Sreekala K

CONTENTS

Preface v

1. **Knitting: Overview** 11
   » *Structure*
   » *Differences between knitting and crocheting*
   » *How to Block Anything*
   » *Reality Knits Episode*
   » *History and culture*
   » *Materials*
   » *Influence on the Industrial Revolution*
2. **The History of Knitting** 51
   » *Origins of knitting*
   » *The history of hand-knitting*
3. **Types of Knitting Needle** 67
   » *Working Patterns in the Round*
   » *Needle sizes and conversions*
   » *Types of hand sewing needles*
4. **Amazing Types of Needlework** 89
   » *Before You Buy A Sewing Machine*
   » *Handspinning woolen yarn*
   » *Different Weights*
   » *Yarn and thread*

5. Understanding Yarn Count **117**

» *Yarn count testing method*

» *Calculating Yarn Count*

» *Indirect system*

» *Most Suitable Count for Knitting Machines*

» *Knitting Machine Parameters*

» *Data Collection*

6. 'Knitting' Growth through Technology and Innovation **127**

» *Knitwear sector going for major automation*

» *Pailung thrives on wide range of knitting machines*

» *Recent Development of Circular Knitting Machine*

» *Loop transfer technology in circular knitting machines*

7. Cable Knitting Patterns **145**

» *What The Back Looks Like*

» *Crochet Pattern: Cabled Headband*

» *Clan Aran patterns*

» *Knitting a standard cable*

» *Checking gauge in a cable pattern*

8. Finger Knitting Technique **163**

» *Uses of Finger Knitting*

» *Modern Practice and Culture*

9. The Art of Hand Knitting **183**

» *Knitting patterns and calculators*

» *How to Finger Knit?*

» *Stitches*

» *Social aspects*

» *Knitting frames (boards,*

» *Knitting Patterns for the Warmest Scarves*

» *Color Knitting Techniques and Methods*

10. Lace Knitting **209**

» *Lace Knitting Technique*

» *Basic knitted fabrics*
» *Double knitting with slip stitches*
» *Mosaic knitting*
» *Knitted Fabric Types*
» *Tricot knitting*

11. Advanced Knitting Techniques **239**

» *Joining Yarn*
» *How Does a Knitting Machine Work?*
» *Knit a Sweater*
» *CHOOSING & PREPARING FABRICS*

Bibliography **257**

Index **259**

1

KNITTING: OVERVIEW

Knitting is a method by which yarn is manipulated to create a textile, or fabric. It is used to create many types of garments. Knitting may be done by hand or by machine. Knitting creates stitches: loops of yarn in a row, either flat or in the round (tubular). There are usually many active stitches on the knitting needle at one time. Knitted fabric consists of a number of consecutive rows of connected loops that intermesh with the next and previous rows. As each row is formed, each newly created loop is pulled through one or more loops from the prior row and placed on the gaining needle so that the loops from the prior row can be pulled off the other needle without unraveling.

Differences in yarn (varying in fibre type, weight, uniformity and twist), needle size, and stitch type allow for a variety of knitted fabrics with different properties, including color, texture, thickness, heat retention, water resistance, and integrity. A small sample of knitwork is known as a swatch.

STRUCTURE

Courses and wales

Like weaving, knitting is a technique for producing a two-dimensional fabric made from a one-dimensional yarn or thread. In weaving, threads are always straight, running parallel either lengthwise (warp threads) or crosswise (weft threads). By contrast, the yarn in knitted fabrics follows a meandering path (a course), forming symmetric loops (also called bights) symmetrically above and below the mean path of the yarn. These meandering loops can be easily stretched in different directions giving knit fabrics much more elasticity than woven fabrics. Depending on the yarn and knitting pattern, knitted garments can stretch as much as 500%. For this reason, knitting was initially developed for garments that must be elastic or stretch in response to the wearer's motions, such as socks and hosiery. For comparison, woven garments stretch mainly along one or other of a related pair of directions that lie roughly diagonally between the warp and the weft, while contracting in the other direction of the pair (stretching and contracting with the bias), and are not very elastic, unless they are woven from stretchable material such as spandex. Knitted garments are often more form-fitting than woven garments, since their elasticity allows them to contour to the body's outline more closely; by contrast, curvature is introduced into most woven garments only with sewn darts, flares, gussets and gores, the seams of which lower the elasticity of the woven fabric still further. Extra curvature can be introduced into knitted garments without seams, as in the heel of a sock; the effect of darts, flares, etc. can be obtained with short rows or by increasing or decreasing the number of stitches. Thread used in weaving is usually much finer than the yarn used in knitting, which can give the knitted fabric more bulk and less drape than a woven fabric.

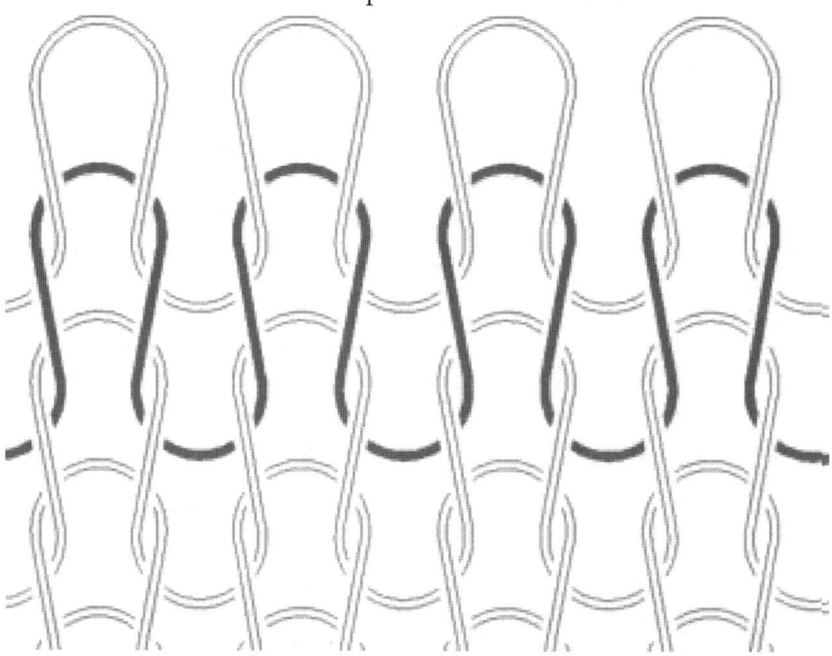

If they are not secured, the loops of a knitted course will come undone when their yarn is pulled; this is known as ripping out, unravelling knitting, or humorously, frogging (because you 'rip it', this sounds like a frog croaking: 'rib-bit'). To secure a stitch, at least one new loop is passed through it. Although the new stitch is itself unsecured ("active" or "live"), it secures the stitch(es) suspended from it. A sequence of stitches in which each stitch is suspended from the next is called a wale. To secure the initial stitches of a knitted fabric, a method for casting on is used; to secure the final stitches in a wale, one uses a method of binding/casting off. During knitting, the active stitches are secured mechanically, either from individual hooks (in knitting machines) or from a knitting needle or frame in hand-knitting.

Weft and warp knitting

There are two major varieties of knitting: weft knitting and warp knitting. In the more common weft knitting, the wales are perpendicular to the course of the yarn. In warp knitting, the wales and courses run roughly parallel. In weft knitting, the entire fabric may be produced from a single yarn, by adding stitches to each wale in turn, moving across the fabric as in a raster scan. By contrast, in warp knitting, one yarn is required for every wale. Since a typical piece of knitted fabric may have hundreds of wales, warp knitting is typically done by machine, whereas weft knitting is done by both hand and machine. Warp-knitted fabrics such as tricot and milanese are resistant to runs, and are commonly used in lingerie.

Weft-knit fabrics may also be knit with multiple yarns, usually to produce interesting color patterns. The two most common approaches are intarsia and stranded colorwork. In intarsia, the yarns are used in well-segregated regions, e.g., a red apple on a field of green; in that case, the yarns are kept on separate spools and only one is knitted at any time. In the more complex stranded approach, two or more yarns alternate repeatedly within one row and all the yarns must be carried along the row, as seen in Fair Isle sweaters. Double knitting can produce two separate knitted fabrics simultaneously (e.g., two socks). However, the two fabrics are usually integrated into one, giving it great warmth and excellent drape.

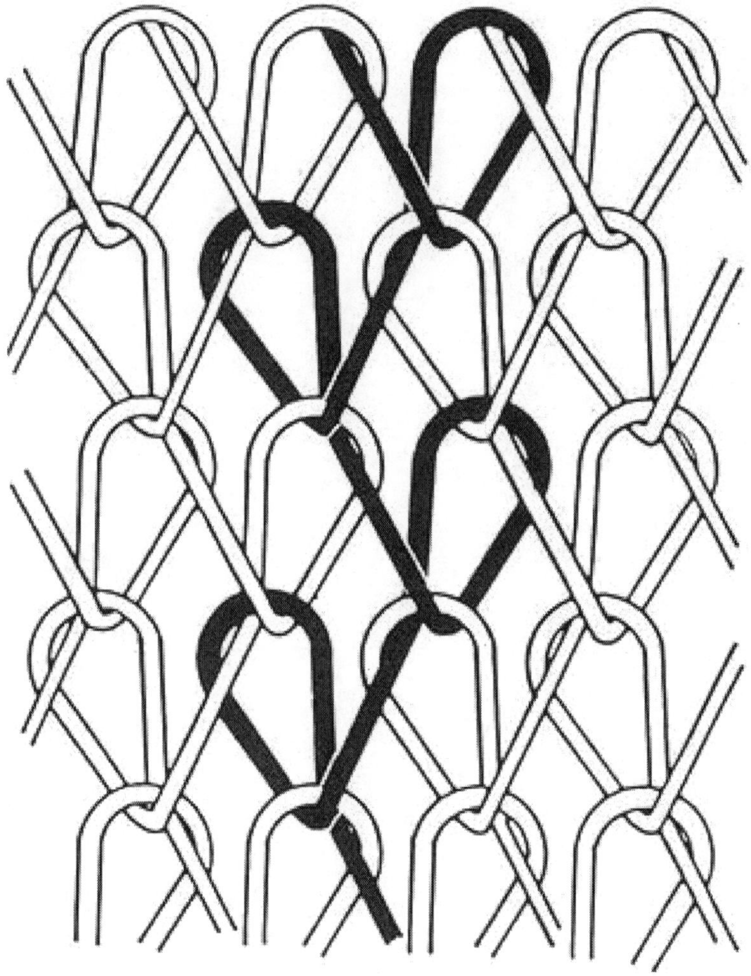

In the knit stitch on the left, the next (red) loop passes through the previous (yellow) loop from below, whereas in the purl stitch (right), the next stitch enters from above. Thus, a knit stitch on one side of the fabric appears as a purl stitch on the other, and vice versa.

Knit and purl stitches

In securing the previous stitch in a wale, the next stitch can pass through the previous loop from either below or above. If the former, the stitch is denoted as a 'knit stitch' or a 'plain stitch;' if the latter, as a 'purl stitch'. The two stitches are related in that a knit stitch seen from one side of the fabric appears as a purl stitch on the other side.

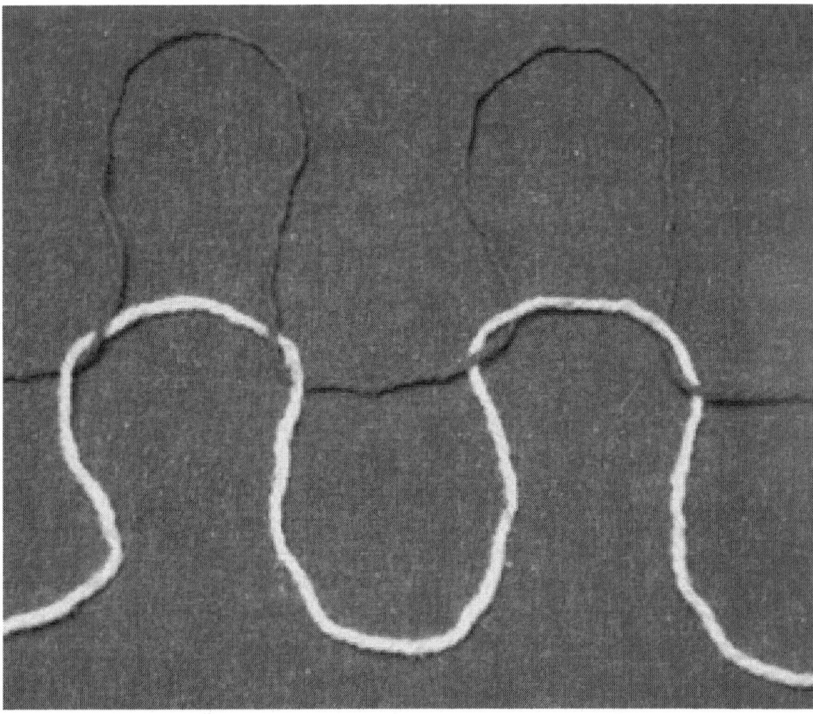

The two types of stitches have a different visual effect; the knit stitches look like 'V's stacked vertically, whereas the purl stitches look like a wavy horizontal line across the fabric. Patterns and pictures can be created in knitted fabrics by using knit and purl stitches as "pixels"; however, such pixels are usually rectangular, rather than square, depending on the gauge/tension of the knitting. Individual stitches, or rows of stitches, may be made taller by drawing more yarn into the new loop (an elongated stitch), which is the basis for uneven knitting: a row of tall stitches may alternate with one or more rows of short stitches for an interesting visual effect. Short and tall stitches may also alternate within a row, forming a fish-like oval pattern.

In the simplest of hand-knitted fabrics, every row of stitches are all knit (or all purl); this creates a garter stitch fabric. Alternating rows of all knit stitches and all purl stitches creates a stockinette pattern/stocking stitch. Vertical stripes (ribbing) are possible by having alternating wales of knit and purl stitches. For example, a common choice is 2x2 ribbing, in which two wales of knit stitches are followed by two wales of purl stitches, etc. Horizontal striping (welting) is also possible, by alternating rows of knit and purl stitches. Checkerboard patterns (basketweave) are also possible, the smallest of which is known as seed/moss stitch: the stitches alternate between knit and purl in every wale and along every row.

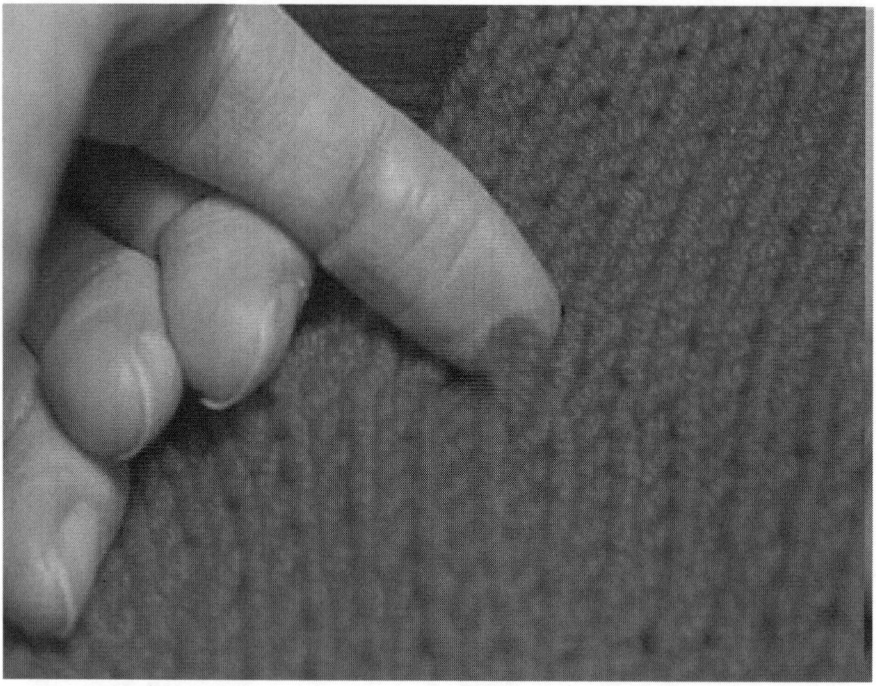

Fabrics in which each knitted row is followed by a purled row, such as in stockinette/stocking stitch, have a tendency to curl—top and bottom curl toward the front (or knitted side) while the sides curl toward the back (or purled side); by contrast, those in which knit and purl stitches are arranged symmetrically (such as ribbing, garter stitch or seed/moss stitch) have more texture and tend to lie flat. Wales of purl stitches have a tendency to recede, whereas those of knit stitches tend to come forward, giving the fabric more stretchability. Thus, the purl wales in ribbing tend to be invisible, since the neighboring knit wales come forward. Conversely, rows of purl stitches tend to form an embossed ridge relative to a row of knit stitches. This is the basis of shadow knitting, in which the appearance of a knitted fabric changes when viewed from different directions.

Typically, a new stitch is passed through a single unsecured ('active') loop, thus lengthening that wale by one stitch. However, this need not be so; the new loop may be passed through an already secured stitch lower down on the fabric, or even between secured stitches (a dip stitch). Depending on the distance between where the loop is drawn through the fabric and where it is knitted, dip stitches can produce a subtle stippling or long lines across the surface of the fabric, e.g., the lower leaves of a flower. The new loop may also be passed between two stitches in the 'present' row, thus clustering the intervening stitches; this approach is often used to produce a smocking effect in the fabric. The new loop may also be passed through 'two or more' previous stitches, producing a decrease and merging wales together. The merged stitches need not be from the same row; for example, a tuck can be formed by knitting stitches together from two different rows, producing a raised horizontal welt on the fabric.

Not every stitch in a row need be knitted; some may be 'missed' (unknitted and passed to the active needle) and knitted on a subsequent row. This is known as slip-stitch knitting. The slipped stitches are naturally longer than the knitted ones. For example, a stitch slipped for one row before knitting would be roughly twice as tall as its knitted counterparts. This can produce interesting visual effects, although the resulting fabric is more rigid because the slipped stitch 'pulls' on its neighbours and is less deformable. Mosaic knitting is a form of slip-stitch knitting that knits alternate colored rows and uses slip stitches to form patterns; mosaic-knit fabrics tend to be stiffer than patterned fabrics produced by other methods such as Fair-Isle knitting.

In some cases, a stitch may be deliberately left unsecured by a new stitch and its wale allowed to disassemble. This is known as drop-stitch knitting, and produces a vertical ladder of see-through holes in the fabric, corresponding to where the wale had been.

DIFFERENCES BETWEEN KNITTING AND CROCHETING

For many beginners, telling the difference between knitting and crocheting is very tricky. Both have the same methods of stitching yarn together, but their methods are fairly different. When knitting, the stitches form a shape that is similar to a "V", while in crochet the stitches are knotted together. Each textile has its own specialties and methods. When knitting, a pair of long needles is needed to be able to form the loops from one set of loop to another through the needles. When crocheting, only one single hook is used to be able to hook the loops together directly onto the clothes. Often times, crocheting is easier to work with at first as compared to knitting.

Right- and left-plaited stitches

Both knit and purl stitches may be twisted: usually once if at all, but sometimes twice and (very rarely) thrice. When seen from above, the twist can be clockwise (right yarn over left) or counterclockwise (left yarn over right); these are denoted as right- and left-plaited stitches, respectively. Hand-knitters generally produce right-plaited stitches by knitting or purling through the back loops, i.e., passing the needle through the initial stitch in an unusual way, but wrapping the yarn as usual.

By contrast, the left-plaited stitch is generally formed by hand-knitters by wrapping the yarn in the opposite way, rather than by any change in the needle. Although they are mirror images in form, right- and left-plaited stitches are functionally equivalent. Both types of plaited stitches give a subtle but interesting visual texture, and tend to draw the fabric inwards, making it stiffer. Plaited stitches are a common method for knitting jewelry from fine metal wire.

Edges and joins between fabrics

The initial and final edges of a knitted fabric are known as the cast-on and bound/cast-off edges. The side edges are known as the selvages; the word derives from "self-edges", meaning that the stitches do not need to be secured by anything else. Many types of selvages have been developed, with different elastic and ornamental properties.

Vertical and horizontal edges can be introduced within a knitted fabric, e.g., for button holes, by binding/casting off and re-casting on again (horizontal) or by knitting the fabrics on either side of a vertical edge separately.

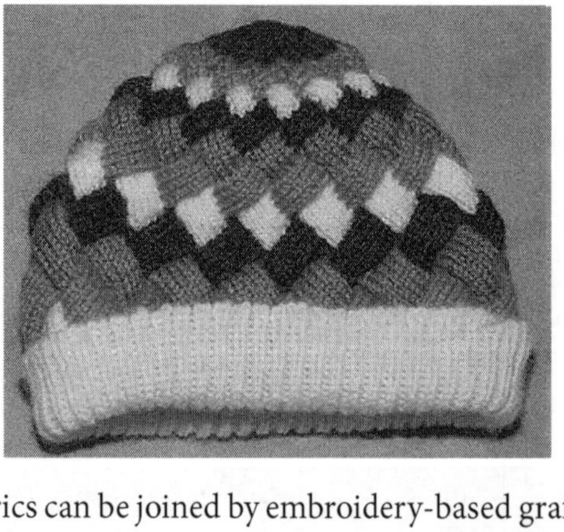

Two knitted fabrics can be joined by embroidery-based grafting methods, most commonly the Kitchener stitch. New wales can be begun from any of the edges of a knitted fabric; this is known as picking up stitches and is the basis for entrelac, in which the wales run perpendicular to one another in a checkerboard pattern.

Cables, increases, and lace

Ordinarily, stitches are knitted in the same order in every row, and the wales of the fabric run parallel and vertically along the fabric. However, this need not be so, since the order in which stitches are knitted may be permuted so that wales cross over one another, forming a cable pattern. Cables patterns tend to draw the fabric together, making it denser and less elastic; Aran sweaters are a common form of knitted cabling. Arbitrarily complex braid patterns can be done in cable knitting, with the proviso that the wales must move ever upwards; it is generally impossible for a wale to move up and then down the fabric. Knitters have developed methods for giving the illusion of a circular wale, such as appear in Celtic knots, but these are inexact approximations. However, such circular wales are possible using Swiss darning, a form of embroidery, or by knitting a tube separately and attaching it to the knitted fabric.

A wale can split into two or more wales using increases, most commonly involving a yarn over. Depending on how the increase is done, there is often a hole in the fabric at the point of the increase. This is used to great effect in lace knitting, which consists of making patterns and pictures using such holes, rather than with the stitches themselves. The large and many holes in lacy knitting makes it extremely elastic; for example, some Shetland "wedding-ring" shawls are so fine that they may be drawn through a wedding ring.

By combining increases and decreases, it is possible to make the direction of a wale slant away from vertical, even in weft knitting. This is the basis for bias knitting, and can be used for visual effect, similar to the direction of a brush-stroke in oil painting.

Ornamentations and additions

Various point-like ornaments may be added to knitting for their look or to improve the wear of the fabric. Examples include various types of bobbles, sequins and beads. Long loops can also be drawn out and secured, forming a "shaggy" texture to the fabric; this is known as loop knitting. Additional patterns can be made on the surface of the knitted fabric using embroidery; if the embroidery resembles knitting, it is often called Swiss darning. Various closures for the garments, such as frogs and buttons can be added; usually buttonholes are knitted into the garment, rather than cut.

Ornamental pieces may also be knitted separately and then attached using applique. For example, differently colored leaves and petals of a flower could be knit separately and attached to form the final picture. Separately knitted tubes can be applied to a knitted fabric to form complex Celtic knots and other patterns that would be difficult to knit.

Unknitted yarns may be worked into knitted fabrics for warmth, as is done in tufting and "weaving" (also known as "couching").

Art of Knitting

Knitting is a method by which thread or yarn is turned into cloth or other fine crafts. Knitted fabric consists of consecutive rows of loops, called stitches. As each row progresses, a new loop is pulled through an existing loop. The active stitches are held on a needle until another loop can be passed through them. This process eventually results in a final product, often a garment.

Knitting may be done by hand or by machine. There exist numerous styles and methods of hand knitting. Different yarns and knitting needles may be used to achieve different end products by giving the final piece a different colour, texture, weight, and/or integrity. Using needles of varying shape and thickness as well as different varieties of yarn can also change the effect.

Knitting the first and last stitch of each row makes your knitting lie flatter and makes it easy to line up secitons when you sew it together. The term "Gauge" refers to how many stitches per inch you are creating horizontally and how many rows you are creating vertically with the needle size and yarn you are using.

To make a garment that is the same size as the pattern you want to make, it is important to match the gauge of the design as given.

After you measure the gauge on the SWATCH you've made, compare your gauge to the gauge given for the pattern you want to make.

If it's the same, you're IN! Knit on!

If it's smaller, go up one needle size.

If it's larger, go down one needle size.

Re-swatch to be sure!

This will save you so... much time.

Try to finish the row you are working on before you set your knitting aside.

Most dropped stitches are from picking up the needles of a work in progress that has been left off in the middle of a row.

Stitches are prone to slide right off the needle tips and might unravel (drop).

Sometimes you won't notice this until later when you have a small hole in the piece.

Like weaving, knitting is a technique for producing a two-dimensional fabric made from a one-dimensional yarn or thread. In weaving, threads are always straight, running parallel either lengthwise (warp threads) or crosswise (weft threads). By contrast, the yarn in knitted fabrics follows a meandering path (a course), forming symmetric loops (also called bights) symmetrically above and below the mean path of the yarn.

These meandering loops can be stretched easily in different directions, which gives knitting much more elasticity than woven fabrics; depending on the yarn and knitting pattern, knitted garments can stretch as much as 500%. For this reason, knitting was initially developed for garments that must be elastic or stretch in response to the wearer's motions, such as socks and hosiery.

For comparison, woven garments stretch mainly along one direction (the bias) and are not very elastic, unless they are woven from stretchable material such as spandex. Knitted garments are often more form-fitting than woven garments, since their elasticity allows them to follow the body's curvature closely; by contrast, curvature is introduced into most woven garments only with sewn darts, flares, gussets and gores, the seams of which lower the elasticity of the woven fabric still further.

Extra curvature can be introduced into knitted garments without seams, as in the heel of a sock; the effect of darts, flares, etc. can be obtained with short rows or by increasing or decreasing the number of stitches. Thread used in weaving is usually much finer than the yarn used in knitting, which can give the knitted fabric more bulk and less drape than a woven fabric.

If they are not secured, the loops of a knitted course will come undone when their yarn is pulled; this is known as ripping out,unravelling knitting, or humorously, frogging (because you 'rip it', this sounds like a frog croaking: 'rib-bit'). To secure a stitch, at least one new loop is passed through it.

Although, the new stitch is itself unsecured ("active" or "live"), it secures the stitch(es) suspended from it. A sequence of stitches in which each stitch is suspended from the next is called a wale. To secure the initial stitches of a knitted fabric, a method for casting on is used; to secure the final stitches in a wale, one uses a method of binding off. During knitting, the active stitches are secured mechanically, either from individual hooks (in knitting machines) or from a knitting needle or frame in hand-knitting.

HOW TO BLOCK ANYTHING

Here's my favorite way to block knitting made from any fiber that can be washed:

- Wash your towels. (!)
- Take a spun-out damp bath towel and smooth it out on a flat surface. Place your sweater pieces on the towel one at a time, smoothing them out and rolling up the towel as you go.
- Let it sit for several hours. Overnight is really good.
- Un-roll your towel and lay out your pieces on a flat surface.
- I use an inexpensive 3'x5' Indian-type rug. The fibers of the rug act like gentle Velcro and grip the sweater pieces, usually without pinning!
- Gently ease your sweater pieces to the size that the pattern called for.
- Let dry and you're all set to put it togeter!
- This method works safely on any fiber content that can be dampened...

Types of Wrap Knitting

Warp knitting comprises several types of knitted fabrics, including tricot, raschel knits, andmilanese knits. All warp-knit fabrics are resistant to runs and relatively easy to sew.

- Tricot is very common in lingerie.
- Milanese is stronger, more stable, smoother and more expensive than tricot and, hence, is used in better lingerie. Milanese is now virtually obsolete.

- Raschel knits do not stretch significantly and are often bulky; consequently, they are often used as an unlined material for coats, jackets, straight skirts and dresses.

The largest outlet for the Raschel Warp Knitting Machine is for lace fabric and trimmings.

How to Find Your Size

If you're having trouble figuring out what size you should make from that new cool pattern you just got (pattern dimensions can be really confusing!) pull out your favorite, best-fitting sweater.

- Measure the width, length, armhole size and sleeve length.
- Compare it to the dimensions given for the pattern you want to make.
- Pick the size closest to your measurements...

Add an Edge Stitch

- Adding an edge stitch to your work neatens the seam when finishing and makes it easy to line up pieces.
- The favorite is to knit the first and last stitch on the right side of the work and purl the first and last stitch on the wrong side.
- You can also knit the first and last stitch on the right side, then slip the first and last stitch purlwise (with yarn in front) on the wrong side. This is especially nice for belts and scarves, making a smooth, flat edge.
- Knitting the first and last stitch of right and wrong side rows makes an edge with a little bump that is decorative and also helps the fabric to lie flat.

How to Choose Needles

Trying a new project using cotton yarn & having trouble keeping your stitches even?

Try using bamboo needles for knitting with cotton!

The slight "tooth" of the bamboo will help you control the loop size of each stitch.

Cotton is slippery and smooth.

Metal needles are great for wool!

Wool is just hairy enough to hold slightly onto the surface of the needle and stays even easier.

After following the pattern for the garter stitch hem, I worked one row of St st on the WS of the work. Now we can start short-rowing on a RS...

1. Ready to short row wrap -we've knit up to the last 3 stitches.
2. Insert right-hand needle into the front of the next stitch (purlwise) and slip onto the right-hand needle.

3. Bring yarn to front of work, then slip stitch back to left needle.
4. Turn work, bring yarn to front (or wherever it needs to be, depending on whether you are knitting or purling), completing the wrap. Finish working the row, short-row wrapping the next indicated stitch.

5. Continue as pattern directs until all short row shaping is finished. Here we see all six stitches wrapped at the side of the garment. Notice how an angle is formed by only partially knitting a segment of the garment.
6. On the final 2 rows, when you knit or purl across, insert needle into wrap first, then knit/purl as appropriate together with the wrapped stitch.

7. This is the bottom - curved hem - that is formed by the technique.Blocking flattens the area thoroughly.
8. Here you can see how the short row shaping forms cuppage for the top of the garment.

That's all there is to it!

After you finish the hem, keep following the pattern to where it refers to the bust shaping.

The pattern says:

"Note: If you wear a C-cup bra or larger, here is the place to put more short rows as worked at the bottom edge. Repeat those instructions once for a C-cup, twice for D, etc. After all short rows have been completed, work one final row across all sts, picking up wraps."

So: for a C-cup you would short row until six stitches on each side of front have been wrapped.

For a D-cup, you would short row until 12 stitches on each side of front have been wrapped.

For a B-cup with ease, you could short row until 3 stitches on each side of front have been wrapped, which would solve the armhole gaping problem many of us have.

The possibilities are endless - think it out. The wonderful thing about this concept is that you can do this on all your designs that have tightness in the front - short row a custom-fit in the area about two-three inches below any armhole shaping.

If you still are having trouble wrapping your brain around this, try this technique out with some scrap yarn first on a swatch using about 30 stitches. Once you get the hang of it, it is like truly a eureka moment. Very easy to do, very tricky to explain.

I-cord is very cool stuff! You can see how to make I-cord as an animation HERE. (Many thanks to DNT-Inc for this and ALL their wonderful animations.)

You use two double pointed needles.

Basically, you knit across your stitches but never turn your work. The right side of your work is always facing you.

To get the yarn to the *working side* or right edge of the next row, you slide your working stitches to the right edge of your double pointed needle, then you simply pull the yarn that's at the end of the row snuggly across the back of your knitting and knit the first stich of the row. Keep doing that and you make a very nice rounded cord. I thought it was truly magic the first time I did it because I'm a skeptical soul and didn't believe it would work. You won't either so just take my word for it and SET YOURSELF FREE!

The Attachment Here is the part that is going to save Rapunzel!I am going to knit one row, then on the next row, I am going to pick up a purl head (a Purl Head is the top of the purl stitch that faces you on the right side of your knitting. I always think of Chicago's homeboy Eddie Vedder when I say or write this because I am a Pearl Jam Head...).I will knit the picked up loop together with the first stitch on my needle.This will attach the I-cord to the sweater with NO PESKY SLIP STITCHING!

Because I am attaching the I-cord to a part of the sweater that is worked in Seed Stitch (K1, P1), the Purl Heads are very obvious and just screaming to be picked up and knit!So I will attach the I-cord every other stitch on the angled bodice. I'm working the piece flat on my lap for ease of view and to help keep the angle *straight*. It's D**n cold here in Chicago, so it's also acting like a cool lap robe fo me :)So here's my technique for Rapunzel:Row 1: Knit all stitches then pull yarn snuggly across back of work and slide all stitches to right edge of needle. Row 2: Pick up a Purl Head from the bottom of the head and K2TOG with the first *real* stitch on your needle.REPEAT these two rows until I-cord is attached around the area you want it to be.

Since my garment was moss stitch, it was easy to attach at every purl head.If you have Stockinette Stitch, just attach into a right or left side of the knit stitch loop, whichever looks best with the design and yarn you are using.

Straight vs. Curved Edges   Depending on the garment design and shape, you might want to attach the I-cord at different intervals than given above. On a

Straight Edge: The rule of thumb is the same as what I use for picking up stitches on a button band: stitch gauge minus 1, ie. if your stitch gauge is 5sts per inch, attach the I-cord to every stitch for 4 stitches, skip a stitch and continue.Every few rows, I slightly stretched the I-cord and checked its position on the garment.

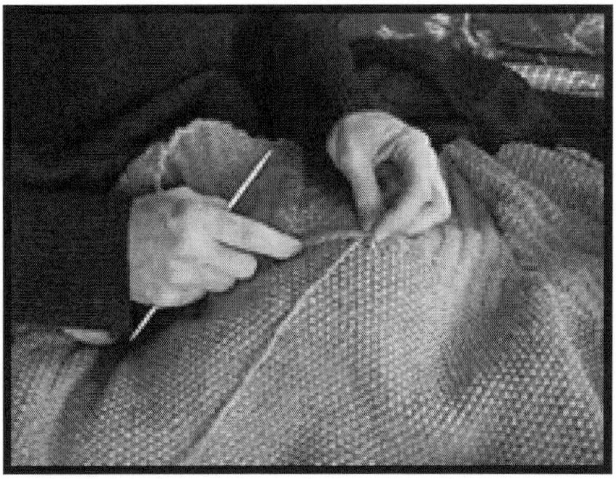

On a Curve or Edge: this is a finagle for sure but on average worsted weight wool it takes 3 stitches to turn a corner on an edge.

The Result Once you start, you can't stop!This is so easy to do and and the results are so pleasing, you'll be applying I-cord all over the place.

Make swirls! Make objects! Make Letters! It's easy.

How to Make a Cable

Making cables is like braiding hair, only you're using two sections instead of three. Once you get the hang of it, you can apply the principle to any number of variations, using different numbers of stitches, different braid directions, or combinations of whole cables to make Aran patterns.

This example uses eight stitches to make a right direction cable. The abbreviation for this is CB4 or Cable Back Four.

Buttons on a Cardigan Pop its Personality

We all love buttons. Many of you have been collecting them at every flea market and place they are found for a long time. When it's time to put that fab creation together, the buttons can make or break the whole concept. Modern? Traditional? Feminine? or Funky, buttons will set the style faster than even the design itself. They are a focal point if you choose a lot of contrast in style or color. Or they can be low-key and fundamental to the design in a subtle way.

REALITY KNITS EPISODE

#1: Whose Size is It Anyway?

Fashion sizing seems to be the 4th dimension. Even Einstein would agree that off-the-rack sizes never actually match commercial knitting patterns with any kind of standards or regularity. Commercial sizes don't even match between price points or manufacturers. It is unreal. Many of us have totally trusted our regular size from a pattern we want to make and find ourselves with armful of awful when we should be jumping for joy after hours of knitting what we thought was going to be a masterpiece. I have had fabulous disasters just blindly following a knitting pattern in *my* size according to the retail clothes I owned. My personal disaster *best* is a sweater I re-knit 3 times! (YES, I am STUBBORN!) AND it still isn't right - the sleeves are too big and the body is tight. Ce la vie!

Exercise #1:Your personal size You need:· fabric tape measure· pen· printout of this picture · a knitting friend

Measure each of the indicated areas and write in the numbers next to the areas of the body on the drawing. Go HERE, a page masterminded by the Craft Council of America, to get information about how you should measure these areas. The info we want is on the right.

It is best if you have a friend help you with this! Not only is it fun, but they will most likely be able to keep the tape measure horizontally aligned better on the parts of you you can't see. And they won't cheat on your hip measurement like I do when I measure myself. Then you can return the favor (and mock themback.)

Exercise #2: Wardrobe Size

Look in your closet and you will see three basic kinds of knitted garments:

- Shells (or vests)
- Pullovers
- Cardigans

They look something like this:

If you look closely, you will see each is differently sized than the other. The Pullover is larger than the Shell. The Cardigan is bigger than both. They all function differently in your wardrobe and need to fit in slightly different ways.

In the design world, this is called *EASE*. It describes how tight the garment fits to your frame. Shells have the least ease or are "very close-fitting". Cardigans have the most ease because they usually go over other clothing or are "standard to loose-fitting". Pullovers are in between; usually we like them to be more "close-fitting".

These great descriptions of general amounts of ease can be found on the same page we found How to Measure in the Fit Chart on the left. The CC of A have gathered this information to try and standardize the knitting world's idea of Ease to be more in line with Retail Clothes. I like it. I THANK THEM! Please print out this page for later. (Better yet, go to the bottom of that page and save their 16 page PDF booklet for your study and pleasure.)

So for this exercise, you need:

- Fabric tape measure
- Pen

- Printout of this picture
- Printout of the Sizing Chart
- Printout of the Fit Chart
- AND, go get from YOUR closet a shell, a pullover, and a cardigan. Pick the ones you think fit you the best.

Measure the areas on these garments that are indicated by a double tipped arrow on the Picture Printout. Write your numbers into the little spaces provided to the nearest half-inch or whole inch. Be sure to multiply the width measurement by 2 so we can compare notes in the next step.

Knit and purl stitches

In securing the previous stitch in a wale, the next stitch can pass through the previous loop from either below or above. If the former, the stitch is denoted as a knit stitch or a plain stitch; if the latter, as a purl stitch. The two stitches are related in that a knit stitch seen from one side of the fabric appears as a purl stitch on the other side.

The two types of stitches have a different visual effect; the knit stitches look like "V"'s stacked vertically, whereas the purl stitches look like a wavy horizontal line across the fabric. Patterns and pictures can be created in knitted fabrics by using knit and purl stitches as "pixels"; however, such pixels are usually rectangular, rather than square, depending on the gauge of the knitting.

Individual stitches, or rows of stitches, may be made taller by drawing more yarn into the new loop (an elongated stitch), which is the basis for uneven knitting: a row of tall stitches may alternate with one or more rows of short stitches for an interesting visual effect. Short and tall stitches may also alternate within a row, forming a fish-like oval pattern.

In the simplest knitted fabrics, all of the stitches are knit or purl; these fabrics are denoted as stockinette and reverse stockinette, respectively. Vertical stripes (ribbing) are possible by having alternating wales of knit and purl stitches; for example, a common choice is 2 x 2 ribbing, in which two wales of knit stitches are followed by two wales of purl stitches, etc.

Horizontal striping (welting) is also possible, by alternating rows of knit and purl stitches; the simplest of these is garter stitch, so-called because its great elasticity made it well-suited for garters. Checker board patterns (basketweave) are also possible, the smallest of which is known as seed stitch: the stitches alternate between knit and purl in every wale and along every row.

Fabrics in which the number of knit and purl stitches are not the same, such as stockinette, have a tendency to curl; by contrast, those in which knit and purl stitches are arranged symmetrically (such as ribbing, garter stitch or seed stitch) tend to lie flat and drape well. Wales of purl stitches have a tendency to recede, whereas those of knit stitches tend to come forward.

Thus, the purl wales in ribbing tend to be invisible, since the neighboring knit wales come forward. Conversely, rows of purl stitches tend to form an embossed ridge relative to a row of knit stitches. This is the basis of shadow knitting, in which the appearance of a knitted fabric changes when viewed from different directions.

Typically, a new stitch is passed through a single unsecured ("active") loop, thus lengthening that wale by one stitch. However, this need not be so; the new loop may be passed through an already secured stitch lower down on the fabric, or even between secured stitches (a dip stitch). Depending on the distance between where the loop is drawn through the fabric and where it is knitted, dip stitches can produce a subtle stippling or long lines across the surface of the fabric, e.g., the lower leaves of a flower.

The new loop may also be passed between two stitches in the present row, thus clustering the intervening stitches; this approach is often used to produce a smocking effect in the fabric. The new loop may also be passed through two or more previous stitches, producing a decrease and merging wales together. The merged stitches need not be from the same row; for example, a tuck can be formed by knitting stitches together from two different rows, producing a raised horizontal welt on the fabric.

Not every stitch in a row need be knitted; some may be left as is and knitted on a subsequent row. This is known as slip-stitch knitting. The slipped stitches are naturally longer than the knitted ones. For example, a stitch slipped for one row before knitting would be roughly twice as tall as its knitted counterparts.

This can produce interesting visual effects, although the resulting fabric is more rigid, because the slipped stitch "pulls" on its neighbours and is less deformable. Slip-stitch knitting plays an important role in mosaic knitting, an important technique in hand-knitting patterned fabrics; mosaic-knit fabrics tend to be stiffer than patterned fabrics produced by other methods such as Fair-Isle knitting.

In some cases, a stitch may be deliberately left unsecured by a new stitch and its wale allowed to disassemble. This is known as drop-stitch knitting, and produces a vertical ladder of see-through holes in the fabric, corresponding to where the wale had been.

Flat knitting versus circular knitting

Circular knitting (also called "knitting in the round") is a form of knitting that can be used to create a seamless tube. Knitting is worked in rounds (the equivalent of rows in flat knitting) in a helix. Originally, circular knitting was done using a set of four or five double-pointed knitting needles. Later, circular needles were invented.

A circular needle resembles two short knitting needles connected by a cable between them. Flat knitting, on the other hand, is used, in its most basic form, to make flat, rectangular pieces of cloth. It is done with two straight knitting needles and is worked in rows, horizontal lines of stitches. A circular knitting needle can also be used to create flat-knitted pieces that are too large for ordinary straight knitting needles, such as afghans and blankets.

Circular knitting is employed to create pieces that are circular or tube-shaped, such as hats,socks, mittens, and sleeves. Flat knitting is usually used to knit flat pieces like scarves,blankets, afghans, and the backs and fronts of sweaters.

There is also such a thing as finger knitting. Instead of needles, fingers are used to produce a tube of knitted fabric.

Felting

Felting is the hand-knitters' term for fulling, a technique for joining knitted or woven wool fibres. The finished product is put in a hot wash and is agitated until it shrinks. While the knitted project is in the washing machine, it is important to check the progress. Depending on the size of the project, some may felt faster than others. The end result typically has reduced dimensions. Bags, mittens, socks and hats are just a few ideas of items that could be felted.

Needle felting

Needle felting is a technique used to add decoration to a knitted or felted piece, where raw roving is applied using a very sharp barbed felting needle by repeatedly piercing the roving and background together. Once washed in hot water, the appliqued decoration is fused with the background. Felted knitting can be cut with scissors without concern about fraying.

HISTORY AND CULTURE

The word is derived from knot, thought to originate from the Dutch verb knutten, which is similar to the Old English cnyttan, to knot. One of the earliest known examples of knitting was cotton socks with stranded knit color patterns, found in Egypt from the end of the first millennium AD. Originally a male-only occupation, the first knitting trade guild was started in Paris in 1527. With the invention of the knitting machine, however, knitting "by hand" became a useful but non-essential craft. Similar toquilting, spinning, and needlepoint, knitting became a leisure activity.

Hand-knitting has gone into and out of fashion many times in the last two centuries, and at the turn of the 21st century it is enjoying a revival. According to the industry group Craft Yarn Council of America, the number of women knitters in the United States age 25-35 increased 150% in the two years between 2002 and 2004. The latest incarnation is less about the "make-do and mend" attitude of the 1940s and early 50s and more about making a statement about individuality as well as developing an innate sense of community.

During the 1940s, English knitting rose in popularity while Continental knitting fell. This is due to the fact that continental knitting originated within Germany and was spread by immigrants. During World War II, continental knitting fell out of style due to its relationship with Germany. It wasn't until Elizabeth Zimmermann publicized continental knitting in the 1980s that it again was popularized.

Additionally, many contemporary knitters have an interest in blogging about their knitting, patterns, and techniques, or joining a virtual community focused on knitting, such as Ravelry, affectionately known as Rav to fiber-lovers around the world. There are also a number of popular knitting podcasts, such as the Manic Purl Podcast and the Savvy Girls Podcast. Contemporary knitting groups may be referred to in the U.S. as a "Stitch 'N Bitch" where a group of knitters get together to work on projects, discuss patterns, troubleshoot their work and just socialize. In the UK, the term has been "knitting circle" since the early 20th century.

Properties of fabrics

The topology of a knitted fabric is relatively complex. Unlike woven fabrics, where strands usually run straight horizontally and vertically, yarn that has been knitted follows a loopy path along its row, as with the red strand in the diagram at left, in which the loops of one row have all been pulled through the loops of the row below it.

Because there is no single straight line of yarn anywhere in the pattern, a knitted piece can stretch in all directions. This elasticity is unavailable from woven fabrics, which only stretch along the bias. Many modern stretchy garments, even as they rely on elastic synthetic materials for some stretch, also achieve at least some of their stretch through knitted patterns.

The basic knitted fabric (as in the diagram, and usually called a stocking or stockinette pattern) has a definite "right side" and "wrong side". On the right side, the visible portions of the loops are the verticals connecting two rows, arranged in a grid ofV shapes. On the wrong side, the ends of the loops are visible, both the tops and bottoms, creating a much more bumpy texture sometimes called reverse stockinette. (Despite being the "wrong side," reverse stockinette is frequently used as a pattern in its own right.) Because the yarn holding rows together is all on the front, and the yarn holding side-by-side stitches together is all on the back, stockinette fabric has a strong tendency to curl toward the front on the top and bottom, and toward the back on the left and right side.

Stitches can be worked from either side, and various patterns are created by mixing regular knit stitches with the "wrong side" stitches, known as purl stitches, either in columns (ribbing), rows (garter, welting), or more complex patterns. Each such fabric has different properties: a garter stitch has much more vertical stretch, while ribbing stretches much more horizontally. Because of their front-back symmetry, these two fabrics have little curl, making them popular as edging, even when their stretch properties are not desired.

Different combinations of knit and purl stitches, along with more advanced techniques, generate fabrics of considerably variable consistency, from gauzy to very dense, from highly stretchy to relatively stiff, from flat to tightly curled, and so on.

Texture

The most common texture for a knitted garment is that generated by the flat stockinette stitch-as seen, though very small, in machine-made stockings and T-shirts-which is worked in the round as nothing but knit stitches, and worked flat as alternating rows of knit and purl. Other simple textures can be made with nothing but knit and purl stitches, including garter stitch, ribbing, and moss and seed stitches. Adding a "slip stitch" (where a loop is passed from one needle to the other) allows for a wide range of textures, including heel and linen stitches, and a number of more complicated patterns.

Some more advanced knitting techniques create a surprising variety of complex textures. Combining certain increases, which can create small eyelet holes in the resulting fabric, with assorted decreases is key to creating knitted lace, a very open fabric resembling lace. Changing the order of stitches from one row to the next, usually with the help of a cable needle or stitch holder, is key to cable knitting, producing an endless variety of cables, honeycombs, ropes, and Aran sweater patterning. Entrelac forms a rich checkerboard texture by knitting small squares,picking up their side edges, and knitting more squares to continue the piece.

The appearance of a garment is also affected by the weight of the yarn, which describes the thickness of the spun fibre. The thicker the yarn, the more visible and apparent stitches will be; the thinner the yarn, the finer the texture.

Color

Plenty of finished knitting projects never use more than a single color of yarn, but there are many ways to work in multiple colors. Some yarns are dyed to be either variegated (changing color every few stitches in a random fashion) or self-striping (changing every few rows). More complicated techniques permit large fields of color (intarsia, for example), busy small-scale patterns of color (such as Fair Isle), or both (double knitting and slip-stitch color, for example).

Yarn with multiple shades of the same hue are called ombre, while a yarn with multiple hues may be known as a given colorway - a green, red and yellow yarn might be dubbed the "Parrot Colorway" by its manufacturer, for example. Heathered yarns contain small amounts of fibre of different colours, while tweed yarns may have greater amounts of different colored fibres.

Process

There are many hundreds of different knitting stitches used by knitters. A piece of knitting begins with the process of casting on (also known as "binding on"), which involves the initial creation of the stitches on the needle. Different methods of casting on are used for different effects: one may be stretchy enough for lace, while another provides a decorative edging -Provisional cast-ons are used when the knitting will continue in both directions from the cast-on.

There are various method employed to "cast on," such as the "thumb method" (also known as "slingshot" or "long-tail" cast-ons), where the stitches are created by a series of loops that will, when knitted, give a very loose edge ideal for "picking up stitches" and knitting a border; the "double needle method" (also known as "knit-on" or "cable cast-on"), whereby each loop placed on the needle is then "knitted on," which produces a firmer edge ideal on its own as a border; and many more. The number of active stitches remains the same as when cast on unless stitches are added (an increase) or removed (a decrease).

Most Western-style knitters follow either the English style (in which the yarn is held in the right hand) or the Continental style (in which the yarn is held in the left hand).

There are also different ways to insert the needle into the stitch. Knitting through the front of a stitch is called Western knitting. Going through the back of a stitch is called Eastern knitting. A third method, called combination knitting, goes through the front of a knit stitch and the back of a purl stitch.

Once the knitted piece is finished, the remaining live stitches are "cast off". Casting (or "binding") off loops the stitches across each other so they can be removed from the needle without unravelling the item. Although the mechanics are different from casting on, there is a similar variety of methods.

In knitting certain articles of clothing, especially larger ones like sweaters, the final knitted garment will be made of several knitted pieces, with individual sections of the garment knit separately and then sewn together. Seamless knitting, where a whole garment is knit as a single piece, is also possible. Elizabeth Zimmermann is probably the best-known proponent of seamless or circular knitting techniques. Smaller items, such as socks and hats, are usually knit in one piece on double-pointed needles or circular needles. (See Circular knitting.)

Mega Knitting

Mega knitting is a term recently coined and relates to the use of knitting needles greater than or equal to half an inch in diameter.

Mega knitting uses the same stitches and techniques as conventional knitting, except that hooks are carved into the ends of the needles. The hooked needles greatly enhance control of the work, catching the stitches and preventing them from slipping off.

It was the development of the knitting machine that introduced hooked needles and enabled faultless, automated knitting. Most knitters probably aren't even aware of the many processes that their fingers perform in the making of a single stitch. However, large gauge needles emphasise those actions and knitting becomes increasingly more awkward when the needle diameter is greater than the width of the knitters finger.

On a one inch diameter (size 50) needle for instance, the shaft begins to taper one and three quarter inches from the tip. This means that the stitches are spread much further apart on mega knitting needles, making them more difficult to control. The hook catches the loop of yarn as each stitch is knitted, meaning that wrists and fingers don't have to work so hard and there is less chance of stitches slipping off the needle. The position of the hook is most important. Turn the left (non-working) hook to face away at all times; turn the right (working) hook toward you up whilst knitting (plain stitch) and away whilst purling.

Mega knitting produces a chunky, bulky fabric or an open lacy weave, depending on the weight and type of yarn used.

MATERIALS

Yarn

Yarn for hand-knitting is usually sold as balls or skeins (hanks), although it may also be wound on spools or cones. Skeins and balls are generally sold with a yarn-band, a label that describes the yarn's weight, length, dye lot, fiber content, washing instructions, suggested needle size, likely gauge, etc. It is common practice to save the yarn band for future reference, especially if additional skeins must be purchased. Knitters generally ensure that the yarn for a project comes from a single dye lot.

The dye lot specifies a group of skeins that were dyed together and thus have precisely the same color; skeins from different dye-lots, even if very similar in color, are usually slightly different and may produce a visible stripe when knitted together. If a knitter buys insufficient yarn of a single dye lot to complete a project, additional skeins of the same dye lot can sometimes be obtained from other yarn stores or online. Otherwise, knitters can alternate skeins every few rows to help the dye lots blend together easier.

The thickness or weight of the yarn is a significant factor in determining the gauge, i.e., how many stitches and rows are required to cover a given area for a given stitch pattern. Thicker yarns generally require thicker knitting needles, whereas thinner yarns may be knit with thick or thin needles. Hence, thicker yarns generally require fewer stitches, and therefore less time, to knit up a given garment.

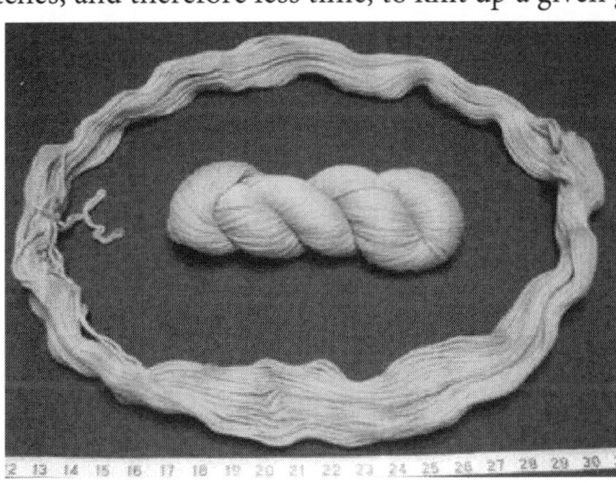

Patterns and motifs are coarser with thicker yarns; thicker yarns produce bold visual effects, whereas thinner yarns are best for refined patterns. Yarns are grouped by thickness into six categories: superfine, fine, light, medium, bulky and superbulky; quantitatively, thickness is measured by the number of wraps per inch (WPI). The related weight per unit length is usually measured in tex or dernier.

Before knitting, the knitter will typically transform a hank into a ball where the yarn emerges from the center of the ball; this making the knitting easier by preventing the yarn from becoming easily tangled. This transformation may be done by hand, or with a device known as a ballwinder. When knitting, some knitters enclose their balls in jars to keep them clean and untangled with other yarns; the free yarn passes through a small hole in the jar-lid.

A yarn's usefulness for a knitting project is judged by several factors, such as its loft (its ability to trap air), its resilience (elasticity under tension), its washability and colorfastness, its hand (its feel, particularly softness vs. scratchiness), its durability against abrasion, its resistance to pilling, itshairiness (fuzziness), its tendency to twist or untwist, its overall weight and drape, its blocking and felting qualities, its comfort (breathability, moisture absorption, wicking properties) and of course its look, which includes its color, sheen, smoothness and ornamental features. Other factors include allergenicity; speed of drying; resistance to chemicals, moths, and mildew; melting point and flammability; retention of static electricity; and the propensity to become stained and to accept dyes.

Different factors may be more significant than others for different knitting projects, so there is no one "best" yarn. The resilience and propensity to (un)twist are general properties that affect the ease of hand-knitting. More resilient yarns are more forgiving of irregularities in tension; highly twisted yarns are sometimes difficult to knit, whereas untwisting yarns can lead to split stitches, in which not all of the yarn is knitted into a stitch. A key factor in knitting is stitch definition, corresponding to how well complicated stitch patterns can be seen when made from a given yarn. Smooth, highly spun yarns are best for showing off stitch patterns; at the other extreme, very fuzzy yarns or eyelash yarns have poor stitch definition, and any complicated stitch pattern would be invisible.

Although knitting may be done with ribbons, metal wire or more exotic filaments, most yarns are made by spinning fibers. In spinning, the fibers are twisted so that the yarn resists breaking under tension; the twisting may be done in either direction, resulting in an Z-twist or S-twist yarn. If the fibers are first aligned by combing them, the yarn is smoother and called a worsted; by contrast, if the fibers are carded but not combed, the yarn is fuzzier and called woolen-spun.

The fibers making up a yarn may be continuous filament fibers such as silk and many synthetics, or they may be staples(fibers of an average length, typically a few inches); naturally filament fibers are sometimes cut up into staples before spinning. The strength of the spun yarn against breaking is determined by the amount of twist, the length of the fibers and the thickness of the yarn. In general, yarns become stronger with more twist (also called worst), longer fibers and thicker yarns (more fibers); for example, thinner yarns require more twist than do thicker yarns to resist breaking under tension. The thickness of the yarn may vary along its length; a slub is a much thicker section in which a mass of fibers is incorporated into the yarn.

The spun fibers are generally divided into animal fibers, plant and synthetic fibers. These fiber types are chemically different, corresponding to proteins, carbohydrates and synthetic polymers, respectively. Animal fibers include silk, but generally are long hairs of animals such as sheep (wool), goat (angora, or cashmere goat),rabbit (angora), llama, alpaca, dog, cat, camel, yak, and muskox (qiviut). Plants used for fibers include cotton, flax (for linen), bamboo,ramie, hemp, jute, nettle, raffia, yucca, coconut husk, banana trees, soy and corn. Rayon and acetate fibers are also produced fromcellulose mainly derived from trees. Common synthetic fibers include acrylics, polyesters such as dacron and ingeo, nylon and other polyamides, and olefins such as polypropylene.

Of these types, wool is generally favored for knitting, chiefly owing to its superiorelasticity, warmth and (sometimes) felting; however, wool is generally less convenient to clean and some people are allergic to it. It is also common to blend different fibers in the yarn, e.g., 85% alpaca and 15% silk. Even within a type of fiber, there can be great variety in the length and thickness of the fibers; for example, Merino wool and Egyptian cotton are favored because they produce exceptionally long, thin (fine) fibers for their type.

A single spun yarn may be knitted as is, or braided or plied with another. In plying, two or more yarns are spun together, almost always in the opposite sense from which they were spun individually; for example, two Z-twist yarns are usually plied with an S-twist. The opposing twist relieves some of the yarns' tendency to curl up and produces a thicker, balanced yarn. Plied yarns may themselves be plied together, producing cabled yarns or multi-stranded yarns. Sometimes, the yarns being plied are fed at different rates, so that one yarn loops around the other, as in bouclé. The single yarns may be dyed separately before plying, or afterwards to give the yarn a uniform look.

The dyeing of yarns is a complex art. Yarns need not be dyed; or they may be dyed one color, or a great variety of colors. Dyeing may be done industrially, by hand or even hand-painted onto the yarn. A great variety of synthetic dyes have been developed since the synthesis of indigo dye in the mid-19th century; however, natural dyes are also possible, although they are generally less brilliant. The color-scheme of a yarn is sometimes called its colorway. Variegated yarns can produce interesting visual effects, such as diagonal stripes; conversely, a variegated yarn may frustrate an otherwise good knitting pattern by producing distasteful color combination.

Tools

The process of knitting has three basic tasks:

1. the active (unsecured) stitches must be held so they don't drop
2. these stitches must be released sometime after they are secured
3. new bights of yarn must be passed through the fabric, usually through active stitches, thus securing them.

In very simple cases, knitting can be done without tools, using only the fingers to do these tasks; however, knitting is usually carried out using tools such as knitting needles, knitting machines or rigid frames. Depending on their size and shape, the rigid frames are called knitting boards, knitting rings (also called knitting looms) or knitting spools (also known as knitting knobbies, knitting nancies, or corkers). There is also a technique of knitting with a crochet hook that has a cord attached to the end, to hold the stitches while they're being worked. This technique is called knooking. Other tools are used to prepare yarn for knitting, to measure and design knitted garments, or to make knitting easier or more comfortable.

Knitting machine

A knitting machine is a device used to create knitted fabrics in a semi- or fully automated fashion. There are numerous types of knitting machines, ranging from the simple, non-mechanical, to the highly complex and electronic. All, however, produce various types of knitted fabrics, usually either flat or tubular, and of varying degrees of complexity. Pattern stitches can be selected by hand manipulation of the needles, or with push-buttons and dials or electronic pattern reading devices and computers.

A few simple devices permit knitting without needles for toy or hobby purposes. The simplest of these is spool knitting, followed by knitting boards or knitting looms, which consist of two rows of pins mounted in two parallel rows approximately 0.5 inches (1.3 cm) apart. Yarn is wound around the pins; various patterns of winding produce different textured knitting.

A needle or special tool is then used to transfer the loops of yarn from around the pins, either off the pins or to other pins, to produce the knitting. Knitting boards can produce complex designs. Other semi-mechanical knitting devices are available.

To produce larger and more complex knitted items such as garments, domestic and industrial machines, with either flat or circular beds that produce rectangular or tubular fabrics are needed. Double bed machines have two flat beds facing each other, in order to produce purl and plain rib fabrics plus a variety of multi patterns. Ribbing attachments can be added to single bed machines to achieve a similar result.

Late 20th century domestic/studio models typically use up to 200 latch-hook needles to hold the stitches in fine, standard, mid-gauge or bulky gauge needle. A carriage or cam box is passed across the bed of needles causing the needle movements required to produce each next stitch. By means of various selection methods, e.g. punch cards, particular needles can be caused to travel by alternate pathways through the cam box.

Thus, needles will knit or not, and the unknitted yarn portions will lie under (slip stitch) or over the needle or be held in the needle hook (tuck stitch). Needles can be placed in holding position to allow short row shaping. In the most modern machines, punchcards have been replaced by computer control.

Automatic patterning machines can knit two-colour Fair Isle patterns automatically, and have machine stitch patterning features such as plating and knitweaving. Plating refers to knitting with two strands of yarn that are held in such a way that one is in front of the other. Plated effects can be particularly striking in a ribbed fabric. Knitweaving refers to a technique in which a separate piece of yarn, often heavier than the knitted fabric, is carried along and caught between stitches to produce an effect like weaving. With knitwoven fabric, the purl side (usually the wrong side) is the right side of the fabric.

The fine and standard gauge models have the option of a lace carriage, where stitches can be transferred from one needle to the next. The yarn passes through a tensioning mechanism and down through the knit carriage, which feeds the yarn to the needles as they knit.

Domestic knitting machines use the weft knitting method which produces a fabric similar to hand knitting. Knitting proceeds more quickly than in hand knitting, where (usually two) straight needles are held in the hand and each stitch is manipulated individually across the row. Knitting machines work an entire row of loops in a single movement.

Comparison to hand knitting

The fabric produced using a knitting machine is of a more even texture than hand-knitted fabric, which is particularly noticeable on large areas of plain stockinette stitch, and can be an advantage. Some stitch patterns (e.g., tuck stitches) are much easier to produce with a knitting machine.

Others (e.g. garter stitch) can also be produced with machine knitting but can take a little longer but still much faster than hand knitting. The standard gauge 200-needle machine can knit the finest yarns up to a good sport-weight, while the heavier yarns knit better on a mid-gauge or bulky knitting machine.

Machine knitting saves a considerable amount of time but does require learning to operate the machines correctly. Most if not all hand knitting patterns can be worked up on a machine, either identically or in a similar design.

Stocking frame

A stocking frame was a mechanical knitting machine used in the textiles industry. It was invented by William Lee of Calverton near Nottingham in 1589. Its use, known traditionally as framework knitting, was the first major stage in the mechanisation of the textile industry, and played an important part in the early history of the Industrial Revolution.

The machine imitated the movements of hand knitters. Lee demonstrated the operation of the device to Queen Elizabeth I, hoping to obtain a patent, but Elizabeth refused, fearing the effects on hand-knitting industries. The original frame had eight needles to the inch, which produced only coarse fabric. Lee later improved the mechanism with 20 needles to the inch.

By 1598, he was able to knit stockings from silk, as well as wool, but was again refused a patent by James I. Lee moved to France with his workers and his machines, but was unable to sustain his business. He died in Paris c.1614. Most of his workers returned to England with their frames, which were sold in London.

The commercial failure of Lee's design might have led to a dead-end for the knitting machine, but John Ashton, one of Lee's assistants, made a crucial improvement by adding the mechanism known as a "divider".

Development

A thriving business built up with the exiled Huguenot silk-spinners who had settled in the village of Spitalfields just outside the city. In 1663, the London Company of Framework Knitters was granted a charter. By about 1785, however, demand was rising for cheaper stockings made of cotton. The frame was adapted but became too expensive for individuals to buy, thus wealthy men bought the machines and hired them out to the knitters, providing the materials and buying the finished product.

With increasing competition, they ignored the standards set by the Chartered Company. In 1728 the Nottingham magistrates refused to accept the authority of the London Company and the centre of the trade moved northwards to Nottingham, which also had a lace making industry.

The breakthrough with cotton stockings, however, came in 1758 when Jedediah Strutt introduced an attachment for the frame which produced what became known as the "Derby Rib". The Nottingham frameworkers found themselves increasingly short of raw materials. Initially they used thread spun in India, but this was expensive and required doubling. Lancashire yarn was spun for fustian and varied in texture.

INFLUENCE ON THE INDUSTRIAL REVOLUTION

It was then that Richard Arkwright arrived with his new experimental spinning machinery. He initially built a works operated by horsepower but it was evident that six to eight would be needed at a time, changed every half hour.

He moved to Cromford and set up what became known as the water frame. Strutt, as his partner, set up mills Belper and Milford. Thus the area joined Nottingham in producing cotton stockings, while Derby, with its mills originated by John Lombe continued largely with silk; Leicester, a farming area, continued with wool.

By 1812, there were estimated to be over 25,000 frames in use, most of them in the three counties, and the frame had come back to Calverton. A legend later developed that Lee had invented the first machine in order to get revenge on a lover who had preferred to concentrate on her knitting rather than attend to him.

A painting illustrating this story was once displayed in the Stocking Framer's Guild hall in London. In 1846, the Victorian artist Alfred Elmore produced a variation on the story in his popular painting The Invention of the Stocking Loom, in which Lee is depicted pondering his idea as he watches his wife knitting (Nottingham Castle Museum).

Techniques for More Successful Knitting

Anyone can learn to knit. Most people learn the knit stitch first, and then the purl stitch. Then they might learn how to cast on and bind off, increase and decrease, cable and seed stitch, and other types of knitting techniques and stitches.

All of these things are knitting techniques, and knitters are constantly learning. With every project there are more and more techniques to learn, such as shaping, seaming, or even steeking, in which you cut your knitwear to make a sweater a cardigan or to open armholes or to insert a heel in a sock.

Knitting excites us. It puts us at ease. Knitting allows us to express our creativity and construct beauty. Knitting challenges us. No matter what drives you to pick up your knitting needles, no matter what your skill level, this free download is a must-have guide that all knitters will appreciate and enjoy.

We made sure this free download, Knitting Techniques for More Successful Knitting, covers basic knitting techniques, such as casting-on and blocking, plus other essential techniques such as short rows and I-cord edging. Knitting can be such a rewarding experience—especially with the use of these helpful knitting techniques in one free download.

Keep a copy in your project bag so you can always have a quick reference for casting on, cable knitting, stitches for embellishing, fixing errant colors, knitting finishing techniques, and more! Expert knitting tips will always be within reach once you download your free eBook.

Wrap-and-Turn Method for Short Rows

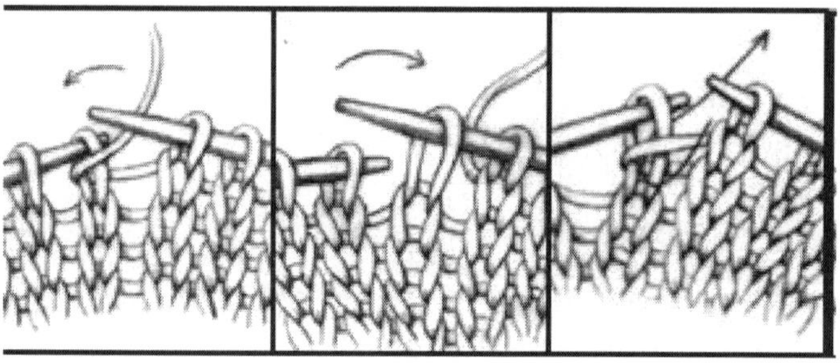

With the knit side facing, knit the required number of stitches to the turning point, slip the next stitch purlwise to the right needle, bring the yarn to the front between the needles, return the slipped stitch to the left needle, bring the working yarn to the back between the needles, and turn the work so that the purl side is facing—one stitch has been wrapped and the yarn is correctly positioned to purl the next stitch. Eventually, you will knit across the wrapped stitch—maybe on the next row or maybe several short-rows later. When you do, hide the wrap (the horizontal bar of yarn across the wrapped stitch) on a knit row as follows: knit to the wrapped stitch, insert the tip of the right needle into both wrap and the wrapped stitch, and knit them together. This forces the turning yarn (the"wrap") to the back (wrong-side) of the fabric.

Tracking Needle Size and Cable Rows

Tracking Your Project Needle Size: Take a very long piece of contrasting scrap yarn and tie the exact number of knots to represent the size of your knitting needle.

If you remove your needle for use on another project, you will always know what size needle you were working with on the unfinished project.

Tracking Needle Size and Tracking Cable Rows

Tracking Your Cable Rows: Tie some scrap yarn with the cast-on yarn tail when working a cable.

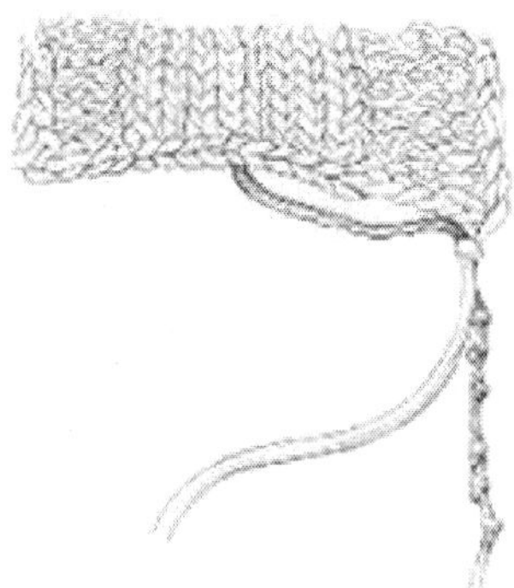

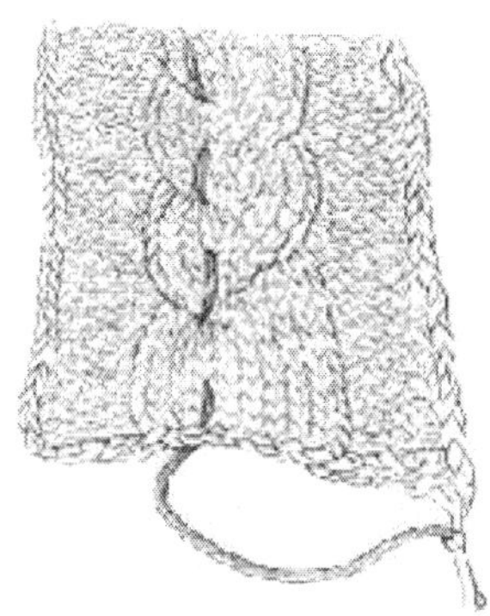

Tracking needle size   Tracking cable rows

Take the scrap yarn and flip it over the needle every four rows—it looks like a running stitch up the project. You will always know what row of the cable you're on as you only have to count four rows.

Why do it every four rows? Because so many cables are based on a multiple of four rows.

Unknitting

Unknitting—sometimes called "tinking" ("tink" is "knit" spelled backward)—is like seeing a reverse. slow-motion film of knitting errors as they happen. Undoing the mistake shows what went wrong, giving us the chance to learn.

Unknitting moves stitches in reverse, undoing each new stitch from the right needle and replacing the old stitch on the left. If you're at the end of a row, leave the full right needle in your right hand, exactly the opposite of what you'd do normally. If you're in the middle of a row, find your working yarn and hold the needle it's attached to in your right hand. The working yarn comes off the back of knit stitches and off the front for purls.

The following illustrations show both unknitting and unpurling; the process is basically the same for both techniques.

How To Unknit, or "Tink"

The trick to unknitting is to place the left needle tip into the stitch lying right underneath the live stitch next to the tip of your right needle. To do this, pull on the working yarn; the opening created in your stitches points to exactly where to place your left needle.

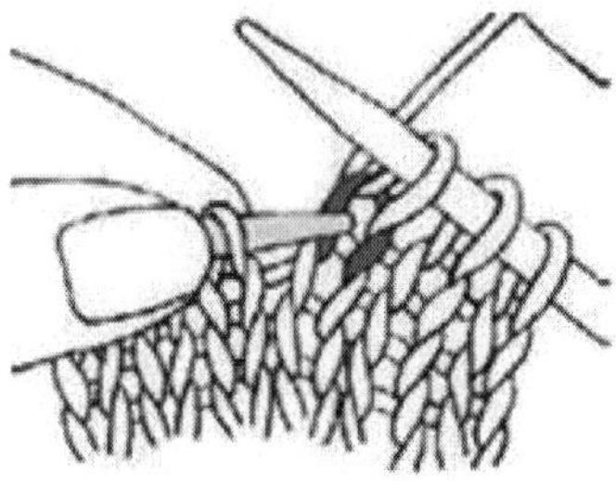

Figure 1.

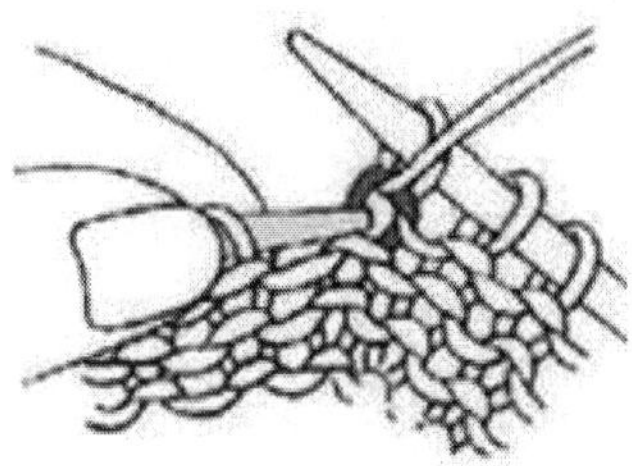

Figure 2.

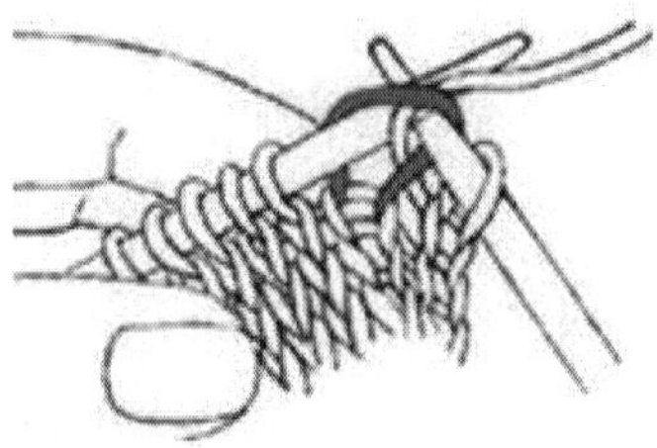

Figure 3.

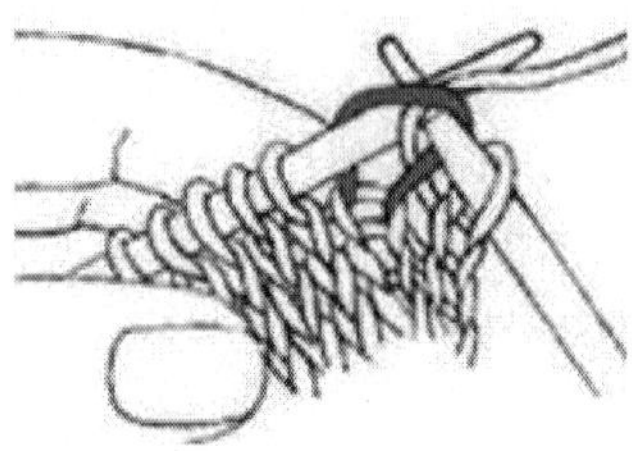

Figure 4.

Place your left needle in the opening from front to back (Figures 1 and 2). This captures the stitch from the previous row without twisting it.

Slip this stitch from right needle to left needle. Pull the working yarn to unravel the stitch. The old stitch now sits on the left needle-one stitch has been unknitted (Above Figures 3 and 4).

Repeat Steps 2 and 3 until you arrive at the mistake, and then fix it!

You'll use this knitting technique all the time.

Blocking Basics

Blocking takes your knitting from good to great. Add the finishing touches using these wet-blocking techniques:

Learn everything you need to know about how to blocking knitting including steam blocking techniques and more!

By definition wet-blocking uses more moisture than steam-blocking, and can be used to stretch and enlarge a knitted piece (although loosely knitted pieces stretch more easily than tightly knitted ones, and any extra inches you gain in width, you may lose in length). There are three degrees of wet-blocking, depending on the amount of moisture added to the knitted fabric.

Spray-blocking

Spray-blocking is the mildest form of wet-blocking. It works equally well for all fibers—although silks and synthetics require more wetness than wool-and it allows for total control over temperature, dampness, and finished texture because you are not restricted to the temperature and amount of steam that comes out of your iron, and you can gently pat and shape the piece with your hands while you work. Pin the handknit to shape right side up on a padded surface placed away from direct sun or heat. Fill a spray bottle with cool tap water and spritz a fine, even mist over the piece. Use your hands to gently pat the moisture into the handknit, if desired, but be careful not to flatten any textured stitches.

Wet-wrapping

Wet-wrapping imparts moisture deeper into the fibers and is appropriate for all types of yarn, especially cotton and acrylic, which are less resilient than wool and require more moisture penetration to reshape stitches. To wet-wrap, thoroughly soak a large bath towel in water, then put it through the spin cycle of a washing machine to remove excess moisture. Place the handknit on top of the towel, then roll the two together jelly-roll fashion. Let the bundle sit until the handknit is completely damp, overnight if necessary. Unroll the towel, remove the handknit, and pin it out to measurements on a padded surface away from direct sun or heat.

Immersion

Immersion imparts moisture thoroughly through the fibers and allows complete reshaping. It is appropriate for all fiber types, and particularly ideal for heavily ribbed or cabled fabrics, or fabrics that have taken on a biased slant during knitting. It is also the method to use after washing a handknit. To immerse a handknit, turn it inside out and soak it in a basin of lukewarm water for about twenty minutes, or until thoroughly wet, gently squeezing water through the piece if necessary. Drain the water, carry the wet handknit in a bundle to the washing machine, and put it through the spin cycle (or roll it in dry towels) to remove excess moisture. Do not twist or wring the handknit. Shape the piece right side up on a padded surface, using pins (and blocking wires) as necessary.

More How-to Block Knitting Tips

Experiment with blocking your gauge swatch before you block an actual knitted piece.

Do not rub, twist, or wring a handknit. Doing so may distort the stitches beyond correction.

Before blocking, weave in all loose ends-the blocking process will help secure the ends in place.

It is preferable to block individual pieces before sewing them together. Blocking makes the sewing process easier and the results of blocking are more consistent when you work with a single layer of fabric. You can block a garment that has been sewed together, but the results may not be as good.

Many experts warn against blocking ribbing, which will lose its natural elasticity if blocked while stretched open. However, ribbing can be successfully blocked if you squeeze it into its most contracted state (so that all the purl stitches recede behind the knit stitches) before you apply moisture.

Allow the blocked handknit to air-dry completely before moving it.

Even More Knitting Techniques for Success

Read through this free eBook to discover simple ways for blocking knitting, fantastic felting ideas, hints for easier cable knitting, finishing techniques, and more. Included in your FREE download are:

Casting on for Ribbing

Durable Cast-Ons

Crochet Chain (Provisional) Cast-On

Embroidery Stitches

Short Rows

Felting Basics

Reverse Single Crochet

I-cord and Applied I-Cord Edging

Three-Needle Bind-Off

Special Trims

There's so much to love about this free download. If you're like most knitters, you're always on the lookout for ways to add durability to your knitting. Well, the cast-on knitting directions found in this free eBook, Knitting Techniques for More

Successful Knitting, will help ensure the durable and stretchy outcome you've been looking for. Looking to jazz up your finished project? Check out the fun embellishment stitches we've included.

Having trouble recalling what size knitting needles you used for that sweater you started last year? There's a simple tip that is sure to make your life so much easier. Can't remember where you left off the last time you were cable knitting? No problem. These easy dilemmas have even easier solutions...you'll wish you had this handy guide years ago! Free knitting help is always nice to have, and now you can refer to the experts whenever you're a little puzzled on what to do next. Improve your knitting skills right away, and add this essential guide to your toolkit today!

2

THE HISTORY OF KNITTING

Knitting is the process of using two or more needles to loop yarn into a series of interconnected loops in order to create a finished garment or some other type of fabric. The word is derived from knot, thought to originate from the Dutch verb knutten, which is similar to the Old English cnyttan, "to knot". Its origins lie in the basic human need for clothing for protection against the elements. More recently, hand knitting has become less a necessary skill and more a hobby.

Hand knitting has been practiced for thousands of years. Where and how this art was invented is still a mystery. Some people believe that knitting began when primitive man made webs out of roots. Others think that Arabian nomads carried the craft into Europe. Still others think that this art originated in Persia. Regardless of where the craft began knitting is the textile infant.

From the beginning the art of knitting was an occupation for women. The oldest remnants of seemingly knitted pieces are those worn as socks. It seems that socks and stockings were knitted because they had to be shaped to the foot or leg. Records seem to indicate that sweaters were first knit in the 17th century. At this time the purl stitch does not seem to have existed. The first time the purl stitch is mentioneddates from the mid 16th century.

The earliest example of true knitting is a pair of knitting socks found in Egypt, dating back to A.D. 1100-just over 9 centuries ago! These early socks were worked in Nalebinding, an ancient craft which used thread to create fabric by making knots and loops. It was done with a wood or bone needle. There are many versions of nalebinding, and some of them look like true knitting.

During Medieval times the craft was controlled by guilds. Knitted garments were worn by the wealthy class. By the 16th century knitting had advanced into a craft. During the Elizabethan era knitting schools were established in Britain. Knitted stockings provided an income for the poor. These stockings were exported to Germany, Holland and Spain. During these times men wore short trunks so fitted stockings where a fashion necessity. Dales knitting, which began at the end of the 16th century, must hold the record for the longest knitting industry in England. Articles from this cottage knitting industry are preserved in the Museum of Hawes in Wensleydale.

Knitting is important in Scottish history. During the 17th and 18th centuries whole families were involved in knitting garments. Sweaters were important to the fisherman of the Scottish Isles. Fair isle and cable patterns were used on aran sweaters. During the French-Nepoleanic wars woman got together to knit socks and mittens for the soldiers. This practice continued through World War I and World War II.

Different types of knitting originated in different parts of the world. Two-color knitting or fair isle knitting began on a group of islands north of Britain. The first example of this intricate pattern is believed to be knitted around 1850. The Prince of Wales wore a fair isle sweater in 1921. Yet other legends say that fair isle knitting was inspired in 1588 when a Spanish ship was wrecked off Fair Isle and the crew inspired native knitters to new forms of knitting patterns. Many lace patterns in use today still have French names. German knitting also has a long history. Four or five needles were often used by German knitters. The picture of the knitting Madonna, The Visit of the Angels, painted around 1390, by a Munich painter, documents this long-established craft of knitting.

The Industrial Revolution took over the hand craft of knitting through the invention of the knitting machine. Hand knitting lost its ability to compete with these knitting machines. Knitting as an art and craft fell to the wayside and was kept alive as a hobby.

It is interesting to watch the renaissance of knitting. Different techniques from different parts of the world are being combined. Specialty yarns have been created from different fibers. These new yarns produce stunning results. Companies are aspiring to make their knitting patterns the height of fashion. Celebrities have been seen knitting, which helps to popularize the revival of the art of knitting. Manufacturers and designers are working together to stimulate interest and creativity within this craft.

ORIGINS OF KNITTING

Knitting is a technique of producing fabric from a strand of yarn or wool. Unlike weaving, knitting does not require a loom or other large equipment, making it a valuable technique for nomadic and non-agrarian peoples. The oldest knitted artifacts are socks from Egypt, dating from the 11th century. They are a very fine gauge, done with complex colourwork and some have a short row heel, which necessitates the purl stitch. These complexities suggest that knitting is even older than the archeological record can prove.

Earlier pieces having a knitted or crocheted appearance have been shown to be made with other techniques, such as Nålebinding, a technique of making fabric by creating multiple loops with a single needle and thread, much like sewing. Some artefacts have a structure so similar to knitting, for example, 3rd-5th century CE Romano-Egyptian toe-socks, that it is thought the "Coptic stitch" of nalbinding is the forerunner to knitting.

Most histories of knitting place its origin somewhere in the Middle East, and from there it spread to Europe by Mediterranean trade routes and later to the Americas with European colonisation.

Knitting is a technique of producing fabric from a strand of yarn or wool. Unlike weaving, knitting does not require a loom or other large equipment, making it a valuable technique for nomadic and non-agrarian peoples.

Early European knitting

The earliest known knitted items in Europe were made by Muslim knitters employed by Spanish Christian royal families. Their high level of knitting skill can be seen in several items found in the tombs in the Abbey of Santa María la Real de Las Huelgas, a royal monastery, near Burgos, Spain. Among them are the knitted cushion covers and gloves found in the tomb of Prince Fernando de la Cerda, who died in 1275. The silk cushion cover was knitted at approximately 20 stitches per inch. It included knitted patterns reflecting the family armoury, as well as the Arabic word baraka ("blessings") in stylized Kufic script. Numerous other knitted garments and accessories, also dating from the mid-13th century, have been found in cathedral treasuries in Spain.

There also is a Votic knitted fragment dated to late 13th century excavated in Estonia. This fragment is knitted in a stranded pattern in three colours and was likely part of a mitten cuff.

Several paintings from Europe portray the Virgin Mary knitting and date from the 14th century, including Our Lady Knitting by Tommaso da Modena (circa 1325-1375) and Visit of the Angel, from the right wing of the Buxtehude Altar, 1400–10, by Master Bertram of Minden.

Archaeological finds from medieval cities all over Europe, such as London, Newcastle, Oslo, Amsterdam and Lübeck, as well as tax lists, prove the spread of knitted goods for everyday use from the 14th century onward. Like many archaeological textiles, most of the finds are only fragments of knitted items so that in most cases their former appearance and use is unknown. One of the exceptions is a 14th or 15th century woollen child's cap from Lübeck.

Although the purl stitch was used in some of the earliest knitted items in Egypt, its knowledge may have been lost in Europe. The first European purl stitches appear in the mid-16th century, in the red silk stockings in which Eleanora de Toledo, wife of Cosimo de Medici, was buried, and which also include the first lacy patterns made by yarn-overs, but the technique may have been developed slightly earlier. The English Queen Elizabeth I herself favoured silk stockings; these were finer, softer, more decorative and much more expensive than those of wool. Stockings reputed to have belonged to her still exist, demonstrating the high quality of the items specifically knitted for her. During this era the manufacture of stockings was of vast importance to many Britons, who knitted with fine wool and exported their wares. Knitting schools were established as a way of providing an income to the poor. The fashion of the period, requiring men to wear short trunks, made fitted stockings a fashion necessity. Stockings made in England were sent to the Netherlands, Spain and Germany.

Many elaborate designs were developed, such as the cable stitch used on Aran sweaters, which was developed in the early 20th century in Ireland.

Importance in Scottish history

Knitting was such an important occupation among those living on the Scottish Isles during the 17th and 18th centuries that whole families were involved in making sweaters, accessories, socks, stockings, etc. Fair Isle techniques were used to create elaborate colourful patterns. Sweaters were essential garments for the fishermen of these islands because the natural oils within the wool provided some element of protection against the harsh weather encountered while out fishing.

Imagine credits to Jodie Morgan.

The oldest knitted artifacts are socks from Egypt, dating from the 11th century CE. They are a very fine gauge, done with complex colorwork and some have a short row heel, which necessitates the purl stitch. These complexities suggest that knitting is even older than the archaeological record can prove.

Earlier pieces having a knitted or crocheted appearance have been shown to be made with other techniques, such as Nålebinding, a technique of making fabric by creating multiple loops with a single needle and thread, much like sewing. Some artifacts have a structure so similar to knitting, for example, 3rd-5th century CE Romano-Egyptian toe-socks, that it is thought the "Coptic stitch" of nalbinding is the forerunner to knitting.

Most histories of knitting place its origin somewhere in the Middle East, and from there it spread to Europe by Mediterranean trade routes and later to the Americas with European colonization.

Early European Knitting:

The earliest known knitted items in Europe were made by Muslim knitters employed by Spanish Christian royal families. Their high level of knitting skill can be seen in several items found in the tombs in the Abbey of Santa María la Real de Las Huelgas, a royal monastery, near Burgos, Spain. Among them are the knitted cushion covers and gloves found in the tomb of Prince Fernando de la Cerda, who died in 1275.

Although the purl stitch is in items from Ancient Egypt, the expertise was probably lost in Europe.

The word knit only became listed in the Oxford Unabridged English Dictionary in the 15th century.

The first examples with the purl stitch in places other than Egypt appeared in the mid 16th century. During the reign of Queen Elizabeth I, the demand for stockings increased. She was a great fan of knitted silk stockings, the fabric was so soft. Knitting schools were set up as a way of providing income for the poor.

Industrial Revolution:

The stocking frame or the mechanical knitting machine was invented in 1589 by William Lee, an English clergyman. After receiving a pair of black stockings from William, Queen Elizabeth I ultimately declined to grant him a patent for his invention. She complained that his machine-made wool stockings were far too coarse for royal ankles. She didn't like the feel of the stockings or their crude form and she was afraid that the machine would take away jobs from her people. However, France's King Henri IV saw the opportunity William's invention provided and offered him financial support. The inventor moved to Rouen where he built a stocking factory. Before long, the French spread the knitting loom throughout Europe. When the device came back to Great Britain, the Worshipful Company of Framework Knitters was incorporated in 1657 in London. Framework knitting was predominantly performed at home, often with the entire family participating.

The city of Nottingham, particularly the district known as Lace Market, was a major producer of machine-knitted lace. Leicestershire and neighbouring counties had long had an association with the hosiery industry. This continued particularly growing with the invention of portable circular knitting machines. Machines could be hired and worked from home rather than relying on a large stocking frame or the much slower hand knitting. One manufacturer of these machines was Griswold, often called Griswold knitting machines, the design of this English sock machine originates from the British inventors, Hainsworth and Griswold.

Some framework knitters were among the Luddites, who resisted the transition to factories. By the middle of the nineteenth century, the knitting industry had still not made the transition to factories. With the improvement of steam-powered knitting machines in the mid-nineteenth century, machine knitting increasingly shifted to factories to accommodate the larger machines.

By the mid-nineteenth century, hand knitting was declining as part of the knitting industry but was increasing as a hobby. Printed knitting patterns and yarn were produced for leisure as well as for industrial use by authors such as Jane Gaugain.

The mechanical knitting machine was invented in 1589 by William Lee, an English clergyman. After receiving a pair of black stockings from William, Queen Elizabeth I ultimately declined to grant him a patent for his invention. She complained that his machine made wool stockings that were far too coarse for royal ankles.

The city of Nottingham, particularly the district known as Lace Market, was a major producer of machine-knitted lace. The city's profits grew with the invention of portable circular knitting machines.

By the middle of the nineteenth century, the knitting industry had still not made the transition to factories. With the improvement of steam-powered knitting machines in the mid-nineteenth century, machine knitting increasingly shifted to factories to accommodate the larger machines.

By the mid-nineteenth century, hand knitting was declining as part of the knitting industry, but was increasing as a hobby. Printed knitting patterns and yarn were produced for leisure as well as for industrial use.

1920s : Including Knits Into Fashion

An increase in the popularity of knitwear was seen in much of the fashion world during this era. Knitwear, especially sweaters/pullovers became essential part of the new fashions of the age for men, women and children, rather than mostly practical garments of associated with particular occupations (e.g., fishermen). Knitwear was often associated with sport and leisure. Garments often became associated with particular sports; for example, white sweaters/pullers, often with coloured stripes (club colours) in the collar, became common for tennis and cricket.

High fashion also embraced knitwear, with Coco Chanel making prominent use of it and Vogue magazine featuring patterns.

Before the 1920s, the majority of commercial knitting in the Western world had centered around production of underwear, socks and hosiery. This vastly expanded as the public taste for knitted fashion did also. Both hand and machine knitting were commercially active on a large scale prior to the Great Depression.

1930s: Depression

The prominence of knitwear in fashion of the 1920s continued. But some changes were seen such as combination of traditional methods and new technologies, they became more common. These changes included addition of zip fasteners and new synthetic yarns in knitwear.

The hardship experienced by many during the Great Depression meant some turned to knitting through necessity. It was much cheaper to knit your own garments than to buy hand (or even machine) knitted products. Knitting was an essential skill during that time as socks, underwear, and other items needed constant repair. Patterns were increasingly featured in women's magazines, reflecting the need for hand-knitting. Some people took a part-time job, hand crafting for profit.

1950s and 60s: Haute Couture

After the war years, knitting had a huge boost as greater colours and styles of yarn were introduced. Many thousands of patterns fed a market hungry for fashionable designs in bright colours. The twinset was an extremely popular combination for the home knitter. It consisted of a short-sleeved top with a long-sleeved cardigan in the same colour, to be worn together.

Girls were taught to knit in school, as it was thought to be a useful skill, not just a hobby. Magazines such as Pins and Needles in the UK carried patterns of varying difficulty including not just clothes, but also blankets, toys, bags, lace curtains and other items that could be sold for profit.

Decline In The 1980s

The popularity of knitting showed a sharp decline during this period in the Western world. Sales of patterns and yarns slumped, as the craft was increasingly seen as old-fashioned and children were rarely taught to knit in school.

The increased availability and low cost of machine-knitted items meant that consumers could have a sweater at the same cost of purchasing the wool and pattern themselves, or often for far less.

Alternatives to traditional woollen knitwear gained in popularity, such as tracksuits and sweatshirts, which began to be worn as everyday wear rather than only in a sporting context.

Advances in technology saw digital versions of knitting machines. As a result, more attention was on the design aspect, rather than uses for fashion or functional means.

1990s

By the late 1980s, many of the suppliers to the home knitting market had disappeared or been absorbed into other companies, while local wool shops suffered a marked reduction in numbers. However, home knitting still had a strong and loyal following.

The growth of craft fairs, release of well researched books on many aspects of knitting and the continued support among those who had learnt the skill in the heyday of the 60s and 70s kept a considerable amount of interest in knitting alive.

One of the most influential changes was the spread of the internet, which enabled knitters to share advice, patterns and experience, but also it meant that home knitters had direct access to supplies rather being reliant on local sources. These trends have continued.

21st Century: Knitwear Gains Popularity

The 21st century has seen a resurgence of knitting. This happened mostly because of the growth of internet and people's interest in DIY crafts.

Natural fibers, from animals such as alpaca, angora, merino and mohair, and plant fibers such as cotton, became more accessible and cheaper to obtain and process. Other fibers, such as bamboo, qiviut, silk, and yak were gaining popularity as well.

The yarn industry began making new types of "Novelty Yarns," using a blend of natural and man-made fibers. These yarns were unique because the final results turned out to be stunning, years of trial and error were not necessary.

Traditional designs blended with non-traditional is happening more than ever today. Many makers hold on to the conventional patterns, which have gained a large following.

This wonderful skill is being adapted, added to, reinvented and blended all the time. What we know as knitting today will be something different in the future.

Nevertheless, it has survived the test of time and will continue to do so.

Origins of Knitting - Making Fabric

Knitting is a technique of making fabric with yarn on two or more needles. The word "knitting" comes from the word "knot". Knitting is made of wool, silk, and other fibers. The first pieces of clothing made in the technique similar to knitting were socks. These socks were done with Nålebinding which is a technique that uses single needle and thread. There are ideas that knitting began in Middle East, from there came to Europe through Mediterranean trade routes and then to the Americas from Europe.

The oldest knitted items have been found in Egypt and are dated between the 11th and 14th centuries AD. Spanish Christian royal families employed Muslim knitters and their works are the earliest known knitted items in Europe. They were very skillful and made many different items like cushion covers and gloves. Archeological findings from the many cities of Europe show that knitting spread throughout Europe in the 14th century. During 16th century, knitting spreads through Britain. On the Scottish Isles, during the 17th and 18th centuries, knitting becomes preoccupation of many. There were cases of whole families who did knitting as a form of a job and it became an important source of their income. Sweaters were one of the major items that were knitted because they proved useful to the local fishermen of these islands. In time, many elaborate designs were developed.

Before Industrial revolution there were mechanical ideas for faster knitting but they didn't spread too far. When the revolution started, machines appeared that did wool spinning, cloth manufacturing and even for knitting lace. During the 1920s, White Russians that lost the Civil War fled to China and showed the craft of knitting to Chinese caravan men, who used camel hair as a material for knitting.

During the Second World War, British wartime government department, the Ministry of Information, started action for people to make knitted items for the Army and Navy to wear in winter. Because wool was in short supply they encouraged people to unpick old unwearable woolen items so they could re-use the wool. Action was also a matter of moral. It showed soldiers that people at home where thinking of them and it gave to people at home a positive sense of contributing to the war effort. After the war knitting again became popular because many new colors appeared along with new designs and patterns for knitting. Children were also thought knitting in schools. In 1980s machine-knitted items became so cheap comparing to handmade that interest for and popularity of hand-knitting started to decline.

With the beginning of the 21st century knitting starts to be popular again. Reason for this popularity is "Handmade Revolution" and ease of finding information on the internet. The more exotic natural fibers from animals, such as alpaca, angora, and merino and plant fibers have become easier and less costly to find in this day as well as designs and patterns thanks, again, to internet. Knitters can also connect easier over the internet and exchange experiences. Another new phenomenon appeared at the same time- yarn bombing which is a type of graffiti where people cover public objects with meters and meters of knitted fabric.

THE HISTORY OF HAND-KNITTING

While the origins of knitting are unclear, we know it has been practised in many different parts of the world, over many centuries, producing objects of great beauty as well as items fulfilling practical needs. When done by hand, it has used simple tools, such as hand-carved sticks of wood, bone, quill and ivory or metal wires and fine steel knitting needles that were commonplace in the 19th century.

The Earliest Examples of Hand-knitting

Hand-knitting today is most closely associated with flat-knitting, which is worked in rows using two knitting needles and where the piece is turned from front to back on each row. But there is also a long history of knitting 'in the round'. Here, four or five needles are used to create a continuous seamless tube – a technique particularly used for caps, stockings and gloves. Today's dominance of flat-knitting is partly because it's easier to represent on construction charts and patterns.

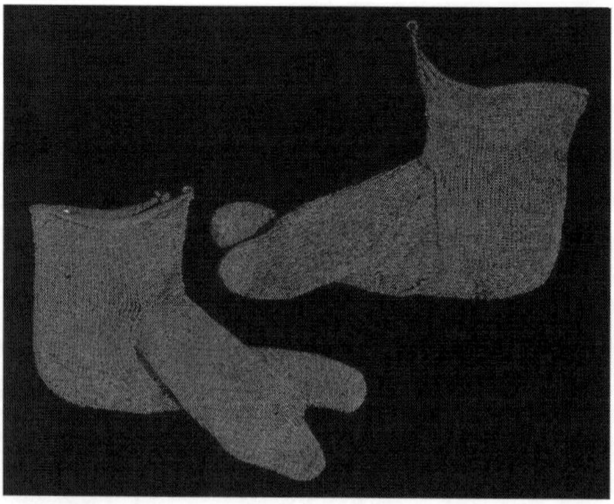

An early woollen item in our collection is a pair of socks from Egypt that date from the 3rd to 5th century AD. They have been made using the technique 'nålbinding', a laborious technique more akin to sewing than knitting, where yarn is threaded through the eye of a sewing needle and then worked in the round through a series of loops. Some believe that this technique was a forerunner of the faster method of knitting with two or more needles.

The earliest example of double-needle knitting in our collection was made in North Africa in about 1100 – 1300, during a period of Islamic rule. The blue and white abstract design echoes the colour combinations and patterning found in Islamic ceramics. It's likely the sock's shaping was achieved by changing the size of the needles during knitting.

The Spread of Knitting in Europe

We know from archaeological finds in medieval cities and surviving tax lists that the use of knitted goods spread in Europe from the 14th century.

In England, the manufacture of knitted caps was sufficiently important to merit control by parliament. The Cappers Act of 1571 stated that every person above the age of six years (except for "Maids, Ladies, Gentlewomen, Noble Personages, and every Lord, Knight and Gentleman of 20 Marks Land") in England on Sundays and holidays should wear (except when travelling), "a Cap of Wool knit, thicked and dressed in England, made within this Realm, and only dressed and finished by some of the Trade of Cappers, upon pain to forfeit for every Day of not wearing three Shillings four Pence". This act helped sustain the production of caps in England, thereby protecting the livelihoods of their makers.

The Museum holds a number of 16th-century knitted caps that were discovered following building works in a house in Worship Street in the City of London. The location of the finds, combined with the fact that the caps are knitted, suggests these caps were made for the middle classes. Many such caps were trimmed with ribbons to imitate more expensive silk versions.

The size of one cap suggests it may have belonged to a young boy. It is closely knitted in wool and was probably once red (possibly dyed in madder), a fashionable colour of the time. These caps were generally constructed in one piece in the round – the method still used for knitted berets today.

Knitted silk jackets were fashionable in the early 17th century, worn as informal dress and known as 'waistcoats'. Our collection includes a particularly fine example of a knitted silk jacket from this period. It was knitted by hand in plain silk yarn and silk partially wrapped in silver thread, in contrasting colours of blue and yellow. The back and two front pieces have been knitted flat before being sewn together, with the abstract floral design imitating the appearance of woven silk textiles. It's been suggested that knitted panels, typically used in the construction of such jackets, were imported from Italy and were put together by the client. The jacket is finely finished with silk-lined sleeves and silver thread buttonholes.

18th-century Knitted Masterpieces

A hand-knitted petticoat in our collection, dating to the first half of the 18th century, demonstrates extraordinary knitting skill, unparalleled in any other known collection. The cream-coloured petticoat has a surface decoration of trees and animals, including an elephant, a lion, an ostrich and a rhinoceros. It has been knitted in the round with no seams and has a circumference of over three metres at the widest point. It's not known exactly how or why a piece of such size and detail was made.

Knitting guilds developed from as early as the 14th century and another virtuoso example of 18th-century knitting in our collection is a carpet or hanging, believed to be an example of a journeyman's work. It took three years of training to become a a trained apprentice or journeyman. To gain full membership of the Hand-Knitters' Guild of Strasbourg, journeymen knitters had to produce 'masterworks' – knitted pieces demonstrating their skill – including a cap, a woollen jacket, a pair of gloves with fingers and a wall hanging patterned with flowers. The Museum's carpet or hanging shows Adam and Eve beneath a central panel depicting Jacob's Dream, from the Old Testament, with a German inscription that translates as "God will continue to send help in despite of mine enemies".

Knitting for Amusement and Education

The stocking frame was the first mechanical knitting machine and was invented by William Lee of Calverton, near Nottingham, in 1589. As its technology developed and the stocking frame became more refined and efficient, factories opened and the rural knitting industry began to die out in Britain. Although it continued in some poorer parts of the country, over the 18th century hand-knitting mainly became the domain of wealthier ladies who had the time to devote to developing the skill. By the mid-19th century, so-called "fancy" knitting was flourishing as an elegant drawing room occupation. Decorative objects such as net purses or pincushions were made by well-to-do women to demonstrate their taste and skill. A pincushion from our collection has been knitted on fine steel needles from cream silk and silver gilt thread. The initials "EW" appear on one side, with the date "1733" on the other, and a decoration of flowers and birds.

As well as being part of a refined lady's repertoire, knitting was deemed an acceptable way for gentlewomen in personal need to earn money. Also seen as a useful skill for poorer members of society, it was taught in orphanages and poor houses. The first recorded knitting schools had been established in Lincoln, Leicester and York in the late 16th century and hand-knitting for income continued in Yorkshire until well into the 19th century. The Ackworth Quaker School in Yorkshire was established in 1779 for girls and boys "not in affluence". According to records, its female pupils knitted 339 stockings in 1821 alone.

The National Society's Instructions on Needlework and Knitting was the first British publication of its type on knitting. It was a handbook for use in the Church of England charity schools run by the National Society, a body seeking to promote Christian education, especially among the poor. The pupils would also have been shown how to sew and knit by a teacher as the instructions in such early books were difficult to follow for anyone unfamiliar with the craft.

19th-Century Celebrations of Hand-Knitting

Hand-knitting featured prominently in the knitwear section of the Great Exhibition, held in Britain in 1851. A number of the submissions were from children and disabled adults and were noted for the remarkable proficiency that they demonstrated. Some of the more unusual entries included a knitted landscape, a shawl knitted with a prayer for the Houses of Parliament and a pair of cuffs, hand-spun and knitted from the wool of French poodles!

Particularly popular were the kind of hand-knitted and beaded lace items that couldn't be copied by knitting machines. The Museum holds an extraordinary baby's gown that won third prize in the Great Exhibition's hand-knitting section. Knitted on extremely fine needles, the gown's patterns of open stitches are similar to those of the knitted Shetland lace shawls popular in the 1840s. It was said that this gown was made with over a million stitches and 6,300 yards of cotton and that its maker, Sarah Ann Cunliffe, worked seven hours a day for five months to complete it.

During the 19th century there was also a boom in needlework publications, with the establishment of series such as Family Friend and Weldon's Practical Needlework. The standard abbreviations now in common use in knitting patterns were established by Weldon's in 1906, while individual pattern leaflets started being printed at the beginning of the 20th century.

Hand-knitting in the 20th Century and Beyond

Home-knitting continued to be a popular way to create items for warmth and for children but it reached another peak during the Second World War when people were encouraged to pick up needles for the war effort. While domestic home-knitting continued through the 1950s, the knitwear made by machine-knitting was more varied than ever and more affordable. It wasn't until the craft countermovement of the 1970s that hand-knitting gained an influence once again.

Designer Kaffe Fassett helped shift attitudes towards hand-knitting in the 1970s. A coat designed by Fassett in our collection reveals Fassett's unorthodox approach to design. It uses a variety of yarns in different thicknesses and textures. Hand-knitted by Zoe Hunt, the order of the stripes and the patterning evolved as the knitting progressed. Fassett's use of geometric patterns and a rich palette of different shades has continued throughout his influential career. He summed up his colour philosophy in a 2002 interview in Vogue Knitting International, saying: "Why limit yourself to just one shade of red when you can use 17?"

Patricia Roberts' hand-knitted cardigan was inspired by the patterns on decorative china plates. The kaleidoscope of stitches in bright primary colours emphasises its handcrafted appeal and the striking patchwork effect is achieved by alternating blocks of different kinds of stitches. In an interview with the V&A, she described her belief that, "the more you invent stitches, the more eloquent you become and the further you can push them". This cardigan is typical of her innovative designs, which use bold colours, contrasting textures and decorative stitches.

3

TYPES OF KNITTING NEEDLE

A knitting needle or knitting pin is a tool in hand-knitting to produce knitted fabrics. They generally have a long shaft and taper at their end, but they are not nearly as sharp as sewing needles. Their purpose is two-fold. The long shaft holds the active (unsecured) stitches of the fabric, to prevent them from unravelling, whereas the tapered ends are used to form new stitches. Most commonly, a new stitch is formed by inserting the tapered end through an active stitch, catching a loop (also called a bight) of fresh yarn and drawing it through the stitch; this secures the initial stitch and forms a new active stitch in its place. In specialized forms of knitting the needle may be passed between active stitches being held on another needle, or indeed between/through inactive stitches that have been knit previously.

The size of a needle is described first by its diameter and secondly by its length. The size of the new stitch is determined in large part by the diameter of the knitting needle used to form it, because that affects the length of the yarn-loop drawn through the previous stitch. Thus, large stitches can be made with large needles, whereas fine knitting requires fine needles. In most cases, the knitting needles being used in hand-knitting are of the same diameter; however, in uneven knitting, needles of different sizes may be used. Larger stitches may also be made by wrapping the yarn more than once around the needles with every stitch. The length of a needle determines how many stitches it can hold at once; for example, very large projects such as a shawl with hundreds of stitches might require a longer needle than a small project such as a scarf or bootie. Various sizing systems for needles are in common use.

Knitting needles come in many styles, sizes, and materials and it can be confusing to know which kind to use when. Whether you 're buying your first set, sorting through a vintage knitting treasure trove, or expanding your collection, this guide is here to help! The diameter or size of knitting needles determines the gauge of your knitting. But beyond this, they also come in different lengths, which affects how big of a project you can work on the needles.

The art of knitting has been in existence for centuries now, and despite the commercialization of the textile industry over the past couple of decades, the art is yet to lose its momentum. In fact, if one might broadly say, the art of knitting has seen a subtle yet gradual transition towards the younger generations as of late.

Knitting is a binding force that keeps the community, and family, connected under one art form. In fact, in some rural households even today, grandmothers, mothers and their daughters spending quality time together knitting.

For newbies venturing into this artform, knitting can be a daunting task. Especially if one is not aware of the basics such as the different types of knitting needles and needle sizes.

The Point

No matter what type of knitting needle you use, the points of the needles can make a big difference. Sharp points make it easier to accurately work with the stitches, especially on fine yarn. Needles with a more rounded point are good for bulky yarn and safer for kids.

Sometimes you'll need a specific type of needle, while other times it's simply a matter of preference. Look at what's available, then find the best option for you!

Getting Started with Circular Needles

The process of casting on is the same with circular needles as it is with straight needles. If you're knitting a round project, the pattern will say something like "join stitches in round, being careful not to twist stitches." There's usually not any explanation of how you actually do that.

There are several different methods for joining a project in the round. Some people simply knit the first stitch very tightly, so that there won't be a gap between the first stitch and the last stitch in the round.

Before I learned to do it the way I do it now, I'd cast on an extra stitch, slip that stitch onto the left-hand needle, and then knit the first stitch and the last stitch together before starting the pattern. That's a fine way to do it, too.

The way I do it now is to slip the last stitch (that is the first one cast on) from the left-hand needle to the right-hand needle, and then lift what's now the second stitch on the right needle (the last one cast on) over the first stitch and onto the left needle. Then pull tight and start knitting with the stitches that are on the left-hand part of the needle.

So what about that "being careful not to twist stitches" part? That's easy to handle as well. Before you make your join, make sure that all the stitches are facing the same direction. That means all the little loopy bits from casting on the stitches should be on the inside of the circle made by the circular needle, without twisting the edge. Then you can make the join and knit as your pattern lays out.

If you don't do this correctly, you'll notice pretty quickly because your knitted tube won't be straight. Unfortunately, the only way to fix a twist like this is to rip it out and start over, so make sure you get it right the first time.

Place a stitch marker on the right-hand needle before you start knitting but after you make the join if you are crossing stitches. This marks the end of the row, which will help you keep track of your pattern.

WORKING PATTERNS IN THE ROUND

When knitting in the round, you are knitting on the right side of the fabric all the time. That means you'll need to alter your basic pattern stitches to get them to come out right.

A lot of circular knitting is done in stockinette stitch, which is great because all you have to do is knit every row. To make garter stitch, instead of knitting every row you'll need to knit one row, purl one row. For reverse stockinette, you purl every row.

Other pattern stitches can be worked in the round, but these are the most common stitches you will come across. Patterns that are designed to be knit in the round usually will be written so you don't have to think about the fact that you're always on the right side, but some will say something like "knit in stockinette (knit every row)" to remind you of what you need to do.

Knitting Flat on Circular Needles

You can also use circular needles to knit something flat. This is especially helpful when working on big projects like afghans, wraps or throws. A circular needle will hold the stitches a little better and make it easier on your body because the needle holds more of the weight of the project.

Knitting flat on circular needles is the same as working on straight needles. Do not join in the round, just cast on and knit. Knit from the left needle to the right as usual, and when you get to the end of the row, switch hands just like you would in knitting with straight needles.

The most commonly-recognized knitting needle shape has one end that is pointed to create new stitches, and the other end has a cap or knob that prevents the stitches from falling off. Single-pointed knitting needles (also known as straight needles) are used in pairs to knit flat pieces.

Circular needles are two straight needles connected with a flexible plastic cord. The cords are available in different lengths (generally from 12" to 60," measured from needle tip to needle tip). These needles can be used for both flat and circular knitting, so some knitters think they are a better investment than single-pointed needles.

If you planning to knit circularly, you will need a needle that is just a little bit smaller in circumference than the item you are knitting. Unless you are planning to use one of the following techniques where circular needles are used for seamless knitting in the round: Magic Loop (long cable needed), 2 Circulars (2 needles of the same size are used) or Travelling Loop (a loop of cable is pulled out as extra).

Double-needed needles (sometimes referred to as "dpns") have points at both ends and are usually sold in sets of four or five. These are used to knit seamless, circular items such as socks. Cable needles are very short and are either straight (like a DPN) or hooked. These serve a specific role in knitting cable stitches.

Interchangeable needle kits are exactly what they sound like—a set of knitting needle tips, and a set of cords of varying length that can attach and detach from the tips to create many different sizes of needles. If you plan to do a lot of knitting, these can be a great investment. A plastic set is available from knitdenise.com, an aluminum set is available in some craft stores under the Boye brand, and knitpicks.com offers both wood and metal sets.

Choose the correct needle size. As knitting needles come in varying diameters and lengths, these factors should be considered when choosing a knitting needles for a project.

- Diameter. The thickness of the needle determines the finished size and look of your finished piece. The thicker the needle, the larger the stitch and the more elastic and loose the resulting fabric. With a smaller the needle, the resulting stitches will be small and the fabric will be usually be tighter and thicker. There are several different sizing systems, but needles are fairly comparable across all systems. Consult the table below for an understanding of sizing systems.
- Length. Although stitches can be squeezed on almost any needle length, choosing one that will allow the stitches to be worked easily from row to row is important. A large project will usually require a longer needle or cord. A small, flat project can be worked on any needle length, but a small circular project will require a small circular needle, double-pointed needles, or a technique used to bypass extra lengths of cord on a long circular needle. Another factor is the yarn chosen for the project (thicker yarn means fewer stitches can fit on a needle). Most straight needles are 10-16 inches (25-40 cm) long, and circular needle cords vary from 12" to as long at 60".

Select the material. Needles are made from many types of materials, the most popular being bamboo, plastic, and aluminum. The knitter's skill level and the yarn being used both need to be taken into consideration when choosing a needle material. Beginners especially need to choose needles that are not so slippery that they will cause the yarn to slip off. A more experienced knitter, on the other hand, may prefer a smooth needle surface to allow the stitches to move easily and quickly.

Bamboo needles. Bamboo is warm to the touch, strong, and tends to be lighter than aluminum. As bamboo has a slight grip, knitted stitches remain in place and bamboo makes for an excellent beginner needle. The material also lends itself to working will with arthritic hands. Bamboo needles are excellent for all types of yarn, especially smoother and lighter weight yarns. Bamboo needles are more flexible than metal, but less than casein or plastic needles. In smaller sizes, however, bamboo can bend or break.

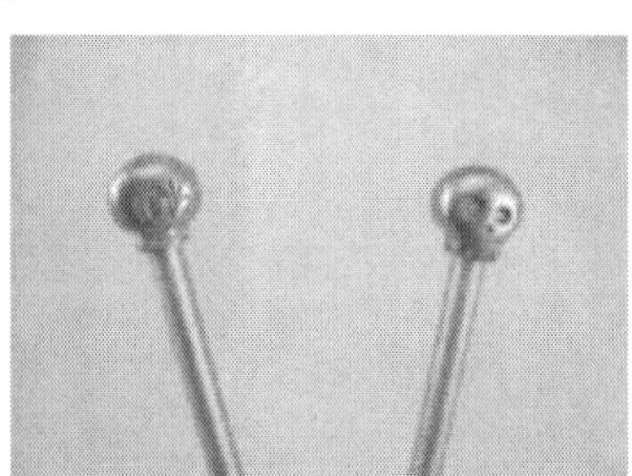

Metal needles are the heaviest needles, but are stronger and do not easily bend.

Metal needles are the heaviest needles, but are stronger and do not easily bend. Metal needles are cold to the touch and slippery. More experienced knitters sometimes prefer them, because the stitches move quickly on the highly polished surface. These are not the best needles for beginning knitters, as stitches can quickly slip off the needle's tip. Metal needles are good for all types of yarn types and weights, especially wool, wool blends, and acrylic. The most common types of metal needles are aluminum, steel, and nickel-plated.

Plastic needles used to be readily available and very popular. They are smooth, sleek, and stitches move quickly. Plastic needles are very flexible. Larger size needles are often made from plastic to reduce their weight. Plastic needles are excellent for all types of yarn and yarn weights.

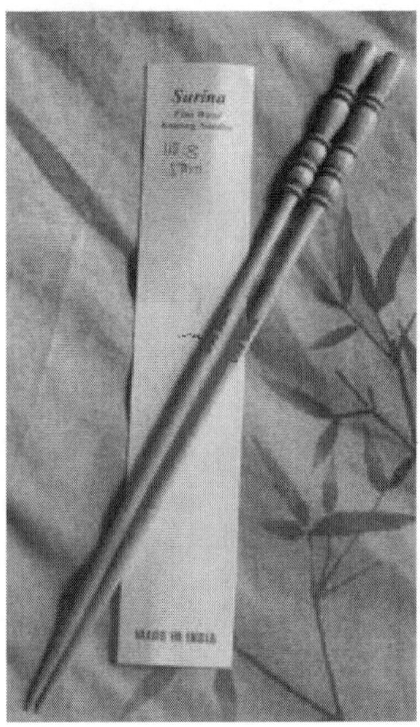

Wooden needles are smooth but not slippery, making them a great choice for beginning knitters and when using slippery yarns.

Wooden needles are smooth but not slippery, making them a great choice for beginning knitters and when using slippery yarns. The finishes are different depending on the brand. These needles come in a variety of woods, the best being made of hardwoods. In very small sizes, like bamboo, wooden needles can break disappointingly easily.

Square needles are said to be particularly good for beginners and those with hand ailments as they make for a more consistent stitch and require less hand tension to maintain in position.

Getting the right gauge. You will need to make a gauge swatch before you start knitting your project. A swatch should be approximately 1 inch larger than the specified gauge as the edges distort the stitches so be prepared to cast on a few extra stitches on each side of the swatch. As stitch size varies from knitter to knitter, the needle size specified in the pattern may not always result in the correct gauge (even when using the yarn in the pattern). So having many needles in varying sizes is an excellent idea.

Knitting needles are a good investment. They can last for a very long time, and can be passed on for generations. Never throw the needles away, unless they are damaged or bent so badly that they cannot be used. If you truly never plan to knit again, give the needles away.

If your yarn has a great deal of texture, you might need a smoother needle, whereas if you use a slippery or a smoother yarn, it is advisable to choose needles that have more grip. Get the feel of the needles before you buy and think of the yarn you are going to use, to determine which will be best.

Examine the points on your needles to see if they are damaged or bent. If so, then it's time to throw them out and buy some new ones.

Keep your needles in a safe place. One good way to store them is to buy or make a knitting needle roll-up. Straight needles can also kept in decorative buckets or pencil cups. Circular needles that do not have the size stamped onto them should be stored in a way that identifies them—often, the bag they came in is suitable for this.

Needles, like socks, have a habit of getting separated, so put a rubber band around matching pairs or sets. However, rubberbands can leave a gummy residue if you use your needles infrequently. Tie them together with yarn or string for longer term storage. Alternatively, you could make or buy a "knitting needle roll-up" similar to the one suggested for straight needles.

- Keep the needles away from small children. They are sharp and if they play with them, someone may get hurt.
- Do not buy all sizes and lengths of needles in one material before you know whether or not you like working with that material. Some knitters love bamboo needles; some knitters feel that it is too fragile, or can snag their work. Know what you like.

A knitting needle or knitting pin is a tool in hand-knitting to produce knitted fabrics. They generally have a long shaft and taper at their end, but they are not nearly as sharp as sewing needles. Their purpose is two-fold. The long shaft holds the active (unsecured) stitches of the fabric, to prevent them from unravelling, whereas the tapered ends are used to form new stitches.

Most commonly, a new stitch is formed by inserting the tapered end through an active stitch, catching a loop (also called a bight) of fresh yarn and drawing it through the stitch; this secures the initial stitch and forms a new active stitch in its place. In specialized forms of knitting the needle may be passed between active stitches being held on another needle, or indeed between/through inactive stitches that have been knit previously.

The size of a needle is described first by its diameter and secondly by its length. The size of the new stitch is determined in large part by the diameter of the knitting needle used to form it, because that affects the length of the yarn-loop drawn through the previous stitch. Thus, large stitches can be made with large needles, whereas fine knitting requires fine needles. In most cases, the knitting needles being used in hand-knitting are of the same diameter; however, in uneven knitting, needles of different sizes may be used.

Larger stitches may also be made by wrapping the yarn more than once around the needles with every stitch. The length of a needle determines how many stitches it can hold at once; for example, very large projects such as a shawl with hundreds of stitches might require a longer needle than a small project such as a scarf or bootie. Various sizing systems for needles are in common use.

Types

Knitting needles in a variety of sizes and materials. Different materials have varying amounts of friction, and are suitable for different yarn types.

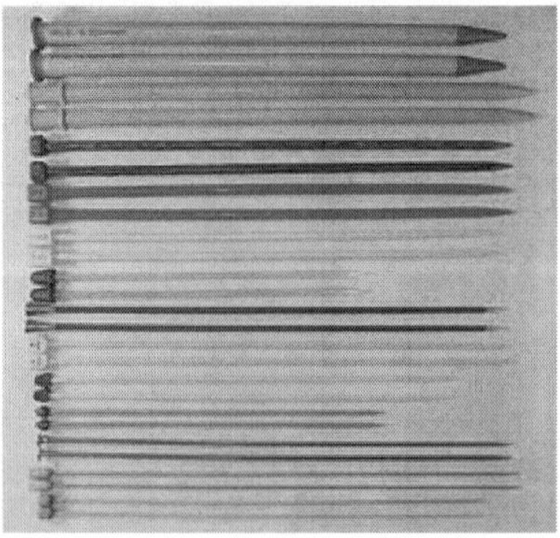

Single-pointed needles

The most widely recognized form of needle is the single-pointed needle. It is a slender, straight stick tapered to a point at one end, with a knob at the other end to prevent stitches from slipping off. Such needles are always used in pairs and are usually 10-16 inches long but, due to the compressibility of knitted fabrics, may be used to knit pieces significantly wider. The knitting of new stitches occurs only at the tapered ends. Fictional depictions of knitting in movies, television programs, animation, and comic strips almost always show knitting done on straight needles. Both Wallace and Gromitand Monty Python, for example, show this type of knitting.

Double-pointed needles

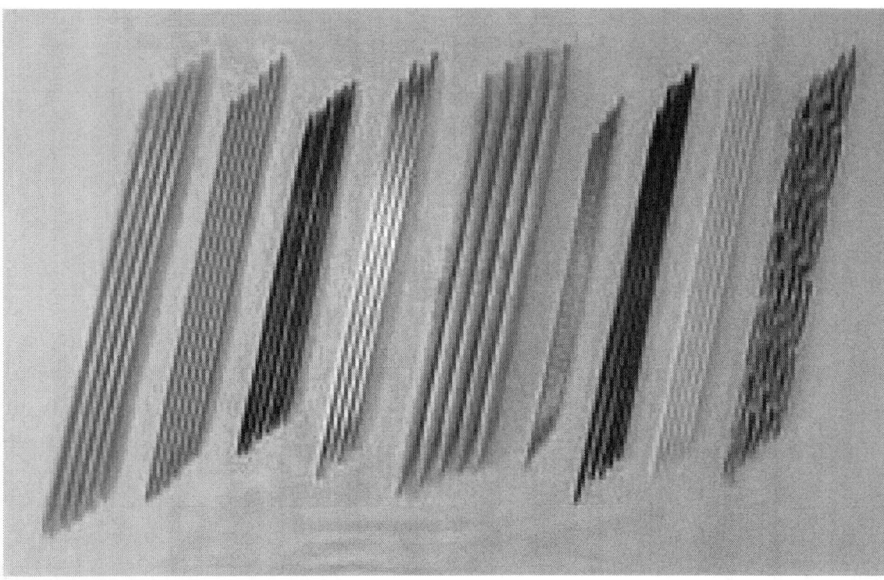

Double-pointed knitting needles in various materials and sizes. They come in sets of four, five or six.

The oldest type of needle is the straight double-pointed needle. Double-pointed needles are tapered at both ends, which allows them to be knit from either end. They are typically used (and sold) in sets of four and five, and are commonly used for circular knitting. Since the invention of the circular needle, they have been most commonly used to knit smaller tube-shaped pieces such as sleeves, collars, and socks. Usually two needles are active while the others hold the remaining stitches. Double-pointed needles are somewhat shorter than single-pointed or circular needles, and are usually around 20 cm in length.

Double-pointed needles are depicted in a number of 14th century oil paintings, typically called Knitting Madonnas, depicting Mary knitting with double-pointed needles (Rutt, 2003).

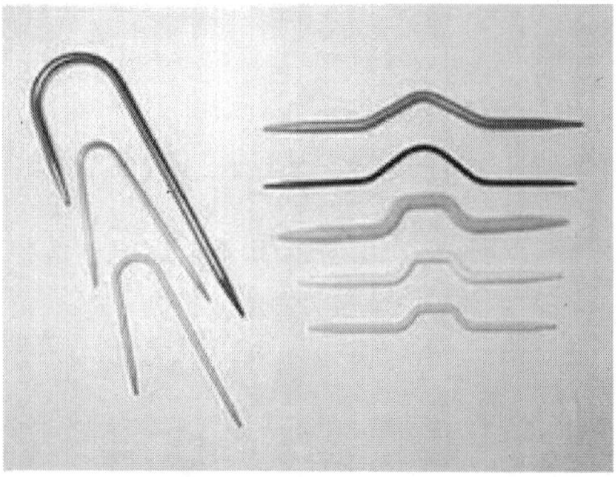

A cable needle is used when knitting cables.

A cable needle is a special type of double-pointed needle that is typically very short and is only used to hold stitches temporarily while the knitter is forming a cable pattern. They are often U-shaped, or have a U-shaped bend, to keep the held stitches from falling off while the primary needle is being used.

Size and Materials

Double-pointed needles range in size from 5 to 8 inches long. Shorter needles work well for tiny projects such as small socks, while the longer ones are good for tubular shaped projects like hats. DPNs come in the same materials as other knitting needles.

Best Uses for Double-Pointed Needles

Most of the time you'll see DPNs in sock knitting, but you can also use them for knitting mittens, gloves, and hats (or just the top of a hat that was started on circular needles). They're also perfect for knitting small toys or even sweater sleeves. Knitting on double-pointed needles takes some practice, but it's worth the effort to have them as an option for small projects knitted in the round.

Circular needles

Circular knitting needles in different lengths, materials and sizes, including plastic, aluminum, steel and nickel-plated brass.

The newest type of needle is the circular needle, a long, flexible double-pointed needle that can be used for both flat and circular knitting. The two tapered ends (typically 5 inches (13 cm) long) are rigid and straight, allowing for easy knitting, and are connected by a flexible strand (usually nylon) that allows the two ends to be brought together. Circular needles can be found in lengths from 9 inches (23 cm) to 40 inches (100 cm). Special kits are available that allow circular needles of various lengths and diameters to be made as needed; rigid ends of various diameters may be screwed into strands of various lengths. The ability to work from either end of one needle is convenient in several types of knitting, such as slip-stitch versions of double knitting. The two ends are used exactly like two needles, in the sense that the knitter holds one in each hand and knits as if having two. The advantages of using circular needles are that the weight of the fabric is more evenly distributed and therefore less taxing on the arms and wrists of the knitter, and also that there is more manoeuvrability of the fabric and needles without fear of the stitches falling off the needles, an occasional occurrence when using double-pointed needles.

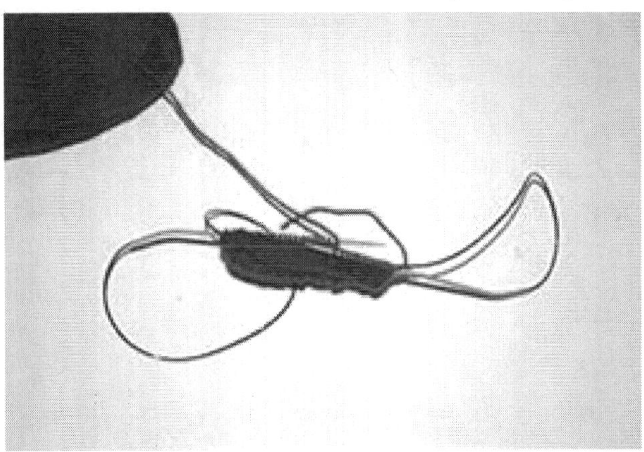

The Magic Loop method may be used to produce narrow tubular items such as socks.

Numerous techniques have been devised for the production of narrow tubular knitting on circular needles. One common method is to use two needles in place of the four or five double-pointed needles traditionally used, while a newer technique is to use one circular needle that is significantly longer than the circumference of the item being knitted. This technique is known as Magic Loop and has recently become a popular method of producing tubular knitting, as only one needle is required.

Julia Hopson and her world record needles

The current holder of this title is Julia Hopson of Penzance in Cornwall. Julia knitted a tension square of ten stitches and ten rows in stocking stitch using knitting needles that were 6.5cm in diameter and 3.5 metres long.

Needle materials

In addition to common wood and metal needles, antique knitting needles were sometimes made fromtortoiseshell, ivory and walrus tusks; these materials are now banned due to their impact on endangered species, and needles made from them are virtually impossible to find.

Modern knitting needles are made of bamboo aluminium, steel, wood, plastic, glass and casein.

Size and Materials

Most circular needles range from 16 to 48 inches long, but they also come in shorter or longer lengths. The cord can also vary in thickness.

As with other needles, steel, aluminum, plastic, bamboo, and wood are the most common materials for the needle part of circular needles.

The cord portion also varies, though nylon and coated steel are popular. Some circular needles hold their shape or have a memory. This can make it awkward to work with the needle because it feels like it has a mind of its own. To avoid this, look for circular needles that say they are memory-free.

Best Uses for Circular Needles

The primary use for circular needles is for projects worked in the round, such as sweaters, hats, cowls, or even socks when using the magic loop method.

You can also use these needles for flat projects. All you have to do is work back and forth, as usual, transferring the work from one end to the other. This is especially helpful for large projects like shawls or blankets which don't often fit on straight needles.

Interchangeable Needles

When you want the flexibility of circular needles in lots of sizes, and the option to use them in different ways, interchangeable needles are a great choice. Like circular needles, they have a firm tip and a flexible cord, but the pieces separate so you can change out the needle sizes and cord lengths. You can also connect needles to cords and attach caps at the ends so you can use them like straight needles.

Interchangeable needles come in sets or as individual pieces. Most connect with a tiny screw-on joining piece, but some use other mechanisms, and each brand only works with its parts.

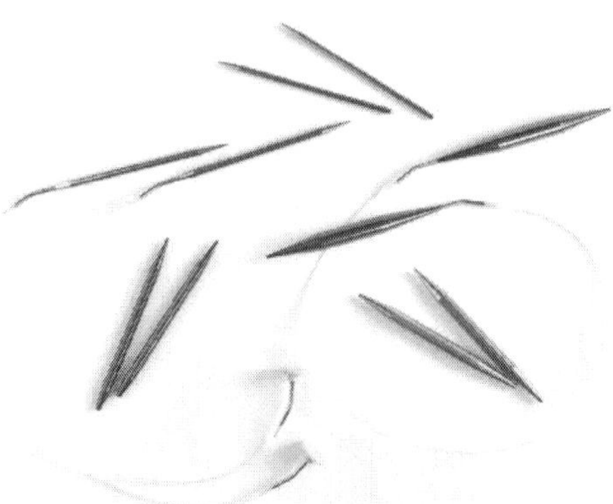

The benefit of buying a set is that you get the pieces you'll need all in one case, although it can be a financial investment. If you want to grow your collection slowly or if you only need a small range of sizes, buying the pieces separately might be the better choice for you.

Needle storage

A tall, cylindrical container with padding on the bottom to keep the points sharp can store straight needles neatly. Fabric or plastic cases similar to cosmetic bags or a chef's knife bag allow straight needles to be stored together but separate, rolled up to maximize space. Circular needles can also be stored this way but are better kept dangling on a hanger device so the cables do not get wound up. If nylon or plastic circular needles are coiled tightly when in storage, it may be necessary to soak them in hot water for a few minutes to get them to uncoil and relax for ease of use.

Needle gauge

A needle gauge makes it possible to determine the size of a knitting needle. Some may also be used to gauge the size of crochet hooks. Most needles come with the size written on the them, but with use and time, the label often wears off, and many needles (like double-pointed needles) tend not to be labelled.

Needle gauges can be made of any material, but are often made of metal and plastic. They tend to be about 3 by 5 inches. There are holes of various sizes through which the needles are passed to determine which hole they fit best, and often a ruler along the edge for determining the gauge of a sample.

NEEDLE SIZES AND CONVERSIONS

Knitting needles with yarn

In the UK, knitting needle 'numbers' are the Standard Wire Gauge designation of the wire from which metal needles are made. The origin of the numbering system is uncertain but it is thought that needle numbers are based on the number of increasingly fine dies that the wire has to be drawn through. In the UK, then, thinner needles thus have a larger number, whereas in the US the opposite applies.

Knitted fabric

Knitted fabrics are the third major class of fabric, after woven and nonwoven fabrics. Knitted fabrics are divided into two basic types: warp-knit fabrics such as tricot and weft-knit fabrics such as a hand-knit sweater. Weft-knit items have the drawback that they run when cut. Warp-knit fabrics are often used in lingerie.

Knitting stitches

Over the long history of knitting across the world, hundreds of different knitting stitches have been created.

The basic building blocks of all hand knitting are the following stitches: knit, purl, cast on, cast off, increase and decrease stitches. Use of a combination of these methods can provide a vast number of different textures to knitted fabrics.

In order to save space in knitting patterns, the names of stitches are normally abbreviated.

Composition of knitted fabrics

The most common fibres used for knitted fabrics are cotton & viscose with or without elastane, these tend to be single jersey construction and are used for most t-shirt style tops.

History of knitwear

Coco Chanel's 1916 use of jersey in her hugely influential suits was a turning point for knitwear, which became associated with the liberated woman. Shortly afterwards, Jean Patou's cubist-inspired, color-blocked knits were the sportswear of choice.

In the 1940s came the iconic wearing of body-skimming sweaters by sex symbols like Lana Turner and Jane Russell, though the 1950s were dominated by conservative popcorn knits. The swinging 1960s were famously manifested in Missoni's colorful zigzag knitwear.This era also saw the rise both of Sonia Rykiel, dubbed the "Queen of Knitwear" for her vibrant striped sweaters and her clingy dresses, and of Kennedy-inspired preppy sweaters.

Sewing needle

This article needs additional citations for verification. Please help improve this article by adding citations to reliable sources. Unsourced material may be challenged and removed.(December 2009)

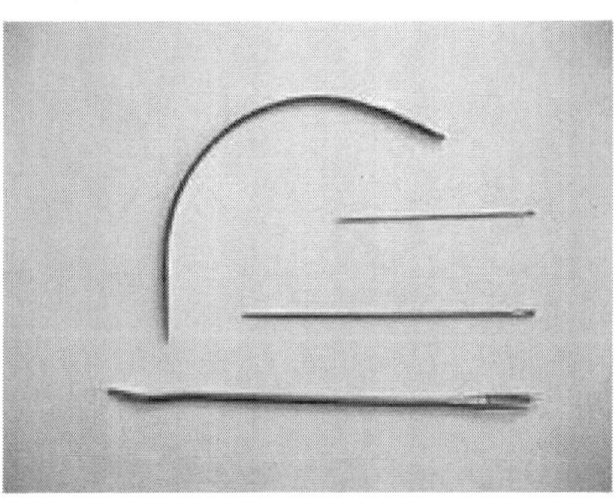

Needles used for hand sewing

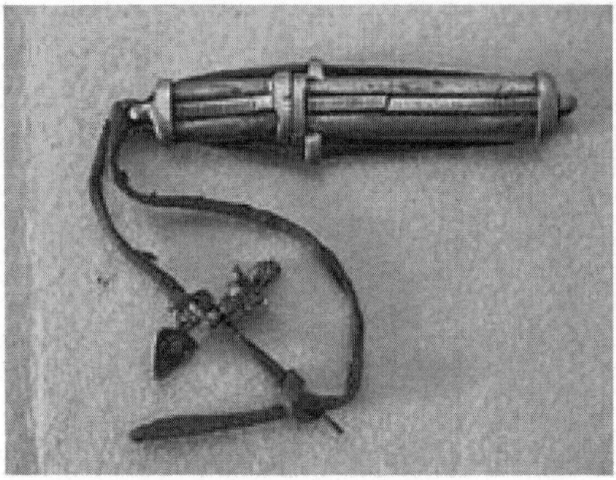

Tibetan needle-case

Eye of a No.5 sharp

Metal container for sewing needles from the second half of the 20th century. From the Museo del Objeto del Objeto collection

A sewing needle is a long slender tool with a pointed tip. The first needles were made of bone or wood; modern ones are manufactured from high carbon steel wire, nickel- or 18K gold plated for corrosion resistance. The highest quality embroidery needles are plated with two-thirds platinum and one-thirds titanium alloy. Traditionally, needles have been kept in needle books or needle cases which have become an object of adornment.

A needle for hand sewing has a hole, called the eye, at the blunt end to carry thread or cord through the fabric after the pointed end pierces it.

TYPES OF HAND SEWING NEEDLES

Hand sewing needles come in a variety of types/ classes designed according to their intended use and in a variety of sizes within each type.

- Sharps: Needles used for general hand sewing; built with a sharp point, a round eye, and are of medium length. Those with a double-eyed head are able to carry two strands of thread while minimizing fabric friction.
- Appliqué: These are considered another all-purpose needle for sewing, appliqué, and patch work.
- Embroidery: Also known as crewel needles; identical to sharps but have a longer eye to enable easier threading of multiple embroidery threads and thicker yarns.
- Betweens or Quilting: These needles are shorter than sharps, with a small rounded eye and are used for making fine stitches on heavy fabrics such as in tailoring, quilt making and other detailed handwork; note that some manufacturers also distinguish between quilting needles and quilting between needles, the latter being slightly shorter and narrower than the former.
- Milliners: A class of needles generally longer than sharps, useful for basting andpleating, normally used in millinery work.
- Easy- or Self-threading: Also called calyxeyed sharps, side threading, and spiral eye needles, these needles have an open slot into which a thread may easily be guided rather than the usual closed eye design.
- Beading: These needles are very fine, with a narrow eye to enable them to fit through the centre of beads and sequins along with a long shaft to thread and hold a number of beads at a time.
- Bodkin: Also called ballpoints, this is a long, thick needle with a ballpoint end and a large, elongated eye. They can be flat or round and are generally used for threading elastic, ribbon or tape through casings and lace openings.

- Chenille: These are similar to tapestry needles but with large, long eyes and a very sharp point to penetrate closely woven fabrics. Useful for ribbon embroidery.
- Darning: Sometimes called finishing needles, these are designed with a blunt tip and large eye making them similar to tapestry needles but longer; yarn darners are the heaviest sub-variety.
- Doll: Not designed for hand sewing at all, these needles are made long and thin and are used for soft sculpturing on dolls, particularly facial details.
- Leather: Also known as glovers and as wedge needles, these have a triangular point designed to pierce leather without tearing it; often used on leather-like materials such as vinyl and plastic.
- Sailmaker: Similar to leather needles, but the triangular point extends further up the shaft; designed for sewing thick canvas or heavy leather.
- Tapestry: The large eye on these needles lets them to carry a heavier weight yarn than other needles, and their blunt tip— usually bent at a slight angle from the rest of the needle— allows them to pass through loosely-woven fabric such as embroidery canvas oreven-weave material without catching or tearing it; comes in a double-eyed version for use on a mounted frame and with two colors of thread.
- Tatting: These are built long with an even thickness for their entire length, including at the eye, to enable thread to be pulled through the double stitches used in tatting.
- Upholstery: These needles are heavy, long needles that may be straight or curved and are used for sewing heavy fabrics, upholstery work, tufting and for tying quilts; the curved variety is practical for difficult situations on furniture where a straight needle will not work Heavy duty 12" needles are used for repairing mattresses. Straight sizes: 3"-12" long, curved: 1.5"-6" long.

Needle size

Needle size is denoted by one or more numbers on the manufacturer's packet. The general convention for sizing of needles, like that of wire gauges, is that within any given class of needle the length and thickness of a needle increases as the size number decreases. For example, a size 9 needle will be thicker and longer than a size 12 needle— however, a size 10 of one class may be (and in some cases actually is) either thinner or finer than a size 12 of another type (see table below). Where a packet contains a needle count followed by two size numbers such as "20 Sharps 5/10" the second set of numbers correspond to the range of sizes of needle within the packet, in this case typically ten sharps needles of size 5 and ten of size 10 (for a total of 20 needles). As another example, a packet labeled "16 Milliners 3/9" would contain 16 milliners needles ranging in sizes from 3 to 9.

Hand Needle Size Chart

Size number	Diameter (mm)	Length	(mm)
Sharps	8	0.61	36.5
Sharps	9	0.61	34.9
Sharps	10	0.46	33.3
Appliqué	9	0.53	34.9
Size number	Diameter (mm)	Length	(mm)
Appliqué	10	0.46	34.3
Appliqué	12	0.46	28.8
Beading	10	0.41	54.6
Beading	13	0.41	48.5
Chenille	18	1.24	48.5
Chenille	20	1.07	45.5
Chenille	22	0.91	40.0
Chenille	24	0.76	37.0
Embroidery	3	0.99	44.5
Embroidery	4	0.91	42.9
Embroidery	5	0.84	41.3
Embroidery	6	0.76	39.7
Embroidery	7	0.69	38.1
Embroidery	8	0.61	36.5
Embroidery	9	0.53	34.9
Milliner	3	0.99	55.6
Milliner	5	0.84	50.8
Milliner	7	0.69	46.8
Milliner	9	0.53	42.9
Quilting needle	8	0.61	28.6
Quilting needle	9	0.56	27.3
Quilting needle	10	0.56	25.8

Quilting needle	12	0.51	22.7
Quilting needle	13	0.51	22.0
Quilting needle between	8	0.61	28.6
Quilting needle between	9	0.53	27.0
Quilting needle between	10	0.46	25.4
Quilting needle between	12	0.53	24.4
Tapestry	18	1.24	48.5
Tapestry	20	1.07	45.5
Size number	Diameter (mm)	Length	(mm)
Tapestry	22	0.91	40.0
Tapestry	24	0.76	37.0
Yarn darners	14	2.01	71.4
Yarn darners	15	1.81	69.9
Yarn darners	16	1.60	68.3
Yarn darners	17	1.40	66.7
Yarn darners	18	1.24	63.5

Needles in archaeology

A variety of archaeological finds illustrate sewing has been present for thousands of years. The Romans left elaborate traces of their sewing technology, especially thimbles and needles. Even earlier Stone Age finds, such as the excavations on the island of Öland at Alby, Sweden, reveal objects such as bone needle cases dating to 6000 BC. Ivory needles were also found dated to 30,000 years ago at the Kostenki site in Russia. The oldest needle in the world was made of bone, dated toAurignacian and discovered in Potok Cave in the Eastern Karavanke,

Slovenia. Native Americans were known to use sewing needles from natural sources. One such source, the agave plant, provided both the needle and the "thread." The agave leaf would be soaked for an extended period of time, leaving a pulp, long, stringy fibers and a sharp tip connecting the ends of the fibers. The "needle" is essentially what was the tip end of the leaf. Once the fibers dried, the fibers and "needle" could then be used to sew together skins and other items used in a cloth-like manner.

Which knitting needle is the best to start with?

All three needles have their own pros and cons, and for an experienced knitter like Sasa, choosing just one favourite is not recommended. It all comes down to the design or pattern, when choosing the right needle to use. If one could say, one needle supplements the other in terms of characteristics.

For instance, though one can knit a sweater using straight needles, the advent of circular needles does make the job easier for a knitter. Now that you're all caught up with the basics of knitting needles and how to use them, its time for you to start your first project. The first step is deciding which project you want to make. It could be muffler, scarf, or beanie and so on.

Next comes the pattern, which is the design or look that you want to go for. Finally, choosing the right needle that compliments the pattern you have chosen. And now you're all set. Get started with your project and don't forget to tag us using our socials as well.

4

AMAZING TYPES OF NEEDLEWORK

Technically needlework is anything done with a needle and thread. The term has come to have a more specialized meaning, however, to separate more elaborate techniques from plain sewing. This chapter discusses such techniques as embroidery, smocking, and quilting.

The term "applique" derives from the Latin word "applico" when means "I apply". Applique is a type of sewing technique where one takes one material and sews it on top of another material with the goal of creating some type of pattern or design. This can be accomplished by hand or machine.

Hand stitching with a needle and thread has existed since ancient times. Most needlework in Greek, Roman, and Chinese civilizations was decorative and symbolic. Fine stitchery with threads of gold and silk showed religious or royal rank. Jewels and beads sewn onto uniforms and embellishing banners indicated great wealth, noble titles, and victories in battle.

In the West the many needlework techniques were expanded and refined as the extravagance of the nobility gave way to the more comfortable clothing and furnishings of the 18th century. The refinement of original techniques and the discovery of new materials led to broad variations in the appearance of needlework. Garments and home furnishings had to serve practical purposes, but embellishment—most often done with needle and thread—added touches of luxury and beauty.

Early settlers in colonial America cherished every scrap of cloth and yarn. American pioneers developed distinctive needlework stitches and methods to conserve fabrics. These early colonists developed skills that used and reused the fabrics woven with their own hands or bought with their frugal savings. Pioneer communities had few opportunities to obtain fabrics from a passing peddler, and even wealthy homemakers could barely afford imported fabrics.

Needlework is decorative sewing and textile arts handicrafts. Anything that uses a needle for construction can be called needlework. Needlework may include related textile crafts such as crochet, worked with a hook, or tatting, worked with a shuttle.

Similar abilities often transfer well between different varieties of needlework, such as fine motor skill and knowledge of textile fibers. Some of the same tools may be used in several different varieties of needlework.

According to the Ladies' Needlework Penny Magazine

There are many women who persuade themselves that the occupations particularly allotted to their sex are extremely frivolous; but it is one of the common errors of a depraved taste to confound simplicity with frivolity. The use of the needle is simple, but not frivolous.

Needlework was an important fact of women's identity during the Victorian age, including embroidery, netting, knitting, crochet, and Berlin wool work. A growing middle class had more leisure time than ever before; printed materials offered homemakers thousands of patterns. Women were still limited to roles in the household, and under the standards of the time a woman working on needle work while entertaining the parlor was considered beautiful. According to one publication from 1843: "Never is beauty and feminine grace so attractive as, when engaged in the honorable discharge of household duties, and domestic cares."

Fancy work was distinguished from plain sewing and it was a mark of a prosperous and well-managed home to display handmade needlework. While plain sewing was often handed over to servants, even in middle class households, fancy work would often be done while entertaining guests, in the afternoons, evenings, or on Sundays. The types of goods that could be decorated with needlework techniques was limited only be the imagination: knitted boots, embroidered book covers, footstools, lampshades, sofa cushions, fans and on and on.

Why choose Plymouth Yarn for your projects? If you knit, crochet or do crafts with yarn, you will find that this brand is incredibly versatile. There are currently 149 yarns to choose from. Let's take a look at some of the products.

The current four best-sellers right now in the Plymouth Yarn Brand Collection are:

Whitney - this is a 91% cotton, 9% nylon Aran weight yarn in solid and variegated colors. While this is a favorite for dishcloths, toys and accessories like purses and bags, you can also make a beautiful summer weight baby afghan that's machine washable and dryable.

Marley - this super bulky weight, variegated yarn is 100% wool and is featured in the Italian Collection. Plymouth Yarnknitting patterns are available for caps, scarves and sweaters that look great with these playful color combinations.

Joy Rainbow Ruffle Yarn - if you really want to have some fun with yarn, try the 85% acrylic, 15% polyester flat yarn available in six solid colors. You'll find a very stylish scarf in Plymouth yarn knitting patterns collection that is perfectly suited to this texture. Plymouth Yarn's bamboo single point knitting needles are a great match for this yarn.

Camina Alpaca from Plymouth Yarn's Alpaca Brand Collection is 98% baby alpaca, 2% nylon Aran weight variegated yarn available in both earth tones and brilliant Peruvian favorites. Alpaca wool has the look and feel of cashmere but is much more durable. It's an excellent alternative for anyone who is allergic to the lanolin in sheep wool.

Some of the other popular yarn combos include Bungee which is 95% Superwash Merino, 5% Elite Spandex; Ashton, a 50% baby alpaca, 40% merino and 10% silk combination from the Bristol Yarn Collection; and 24K, a sparkly 83% nylon, 18% lame Worsted weight yarn that shines when you're planning holiday projects.

If you haven't checked out Plymouth Yarn needles collection, you're missing some wonderful tools for your projects. There are single point bamboo needles as well as bamboo circular needles in sizes 3 to 17. But the real treat is the Rosewood Interchangeable Needle Set. There are 9 pairs of rosewood needle tips in a faux leather carrying case and five color-coded cables with cable caps so that you can move between projects with ease. If you like to take your knitting with you when you travel, this is the perfect accessory. The Transportation Security Administration will confiscate your metal knitting needles, but they allow non-metal circular needles in flight (be sure to carry blunt point scissors).

Needlework is a broad term for the handicrafts of decorative stitching and textile arts. Anything that uses needle for construction is known as needlework. Needle work is exclusive; it has lovely names due to its associations with stunning aspects of life and also the stunning styles.

1.Chikan Embroidery

The craft of Chikan work, usually cited as Lucknow Chicken, is over 400 years old with a firm presence within the Indian and world fashion arena. The technique of its creation is named Chicken and its distinctive sensibility flaunts grace and magnificence as subtly as the user pleases.

While the word Chicken quite virtually suggests that embroidery, the variety incorporates about thirty six totally different sewing techniques. Whereas its central hub and place of origin is Lucknow. Chicken work has unfold so much and wide inside Republic of India, with province and Awadh additionally specializing in its production. In chikan, the needle is control within the right hand while inserting it into the material; the left hand supports and controls the thread so the stitches take the correct form. In ancient chikan no frames were used (though they're used now); the portion of the material to be embroidered is placed over the index of the hand, supported by the rest of the fingers, leaving the thumb free.

The needle is force away from the needle worker who starts from the closest finish and finished at the farthest finish. Get acquainted with Indian Chikan embroidery work in the following:

1. Chikan salwar kameez
2. Chikan Shirts and Kurtas
3. Chikan Home Furnishings
4. Chikan saree

Chikan salwar kameez

Chikan trousers kameez are available in form of colors, styles and patterns. Form of chikankari works is available currently like murri, shadow and phanda, tepchi and daraz etc. Chicken salwar tunic is on the market with significant embroidered and light embroidered provides a very sleek look and can be worn according to occasions. Unstitched trousers suits and Kurtas are on the market in varied colours and materials which can be worn with an identical churidar.

Chikan decorated Shirts and Kurtas

chikan decorated kurtas and shirts are specially designed for men which can be worn in all kinds of occasions. Specially appropriate for summer, these chikan shirts and kurtas are available in fine cotton material with ancient chikan embroidery of various varieties like Aipchi, Bakhia, Phunda, Murri, Jaali, Hathkati, Pechni, Ghas Patti, Chaana Patti etc. whatever the style, length, design, color, there's one thing in chikan embroidery that provides a sleek and stylish look to anyone who wears any dress with chikan decorated work on that.

Chikan Home Furnishings

Today chikan embroidery has become so in style that it's extended to home furnishings also. There are certain exclusive creations using Chikan add home furnishings like Cushion covers, bed Covers, bed Sheets, table material, pillow covers,napkins,curtains,place Mats The fabric used for home furnishings are cotton, mulmul, voile, fabric and even silk. Chikan work home furnishings is ethnic, ancient and yet fashionable in its look.

Chikan saree

Saree is each Indian woman's desire of excellent attire, which stays forever within the fashion scene and provides a classy, feminine, sleek and stylish look to any woman. Chikan dress offers a really pure quiet comfort and might be worn in festal occasions likewise as a daily wear. It's on the market altogether shades of colors however the foremost favored ones are the light pastel shade colors. Chikan sarees in fabric, organza, polyester, chiffon, viscose, georgette, polyester fabric, cotton crepe also are on the market.

2. Crewel Embroidery

Crewel embroidery is a form of surface embroidery using wool. Crewel work is crewel work because it's worked in crewel wool that could be a strong, long thread spun from wool. Crewel embroidery, the materials utilized in crewel work and the forms of stitches common to the present fashionable type of hand embroidery. Style of embroidery threads appropriate for crewel work on the market on the market, from ancient Appleton wool to exceptionally sleek domestic sheep wools that are a delight to sew with, to hand-spun wools artificial with natural vegetable dyes, to wool blends mixed with different fibers like silk, alpaca, mohair, and therefore the like. Traditionally, crewel embroidery is worked on linen, either in a weave or a daily weave. In trendy crewel work, materials like wool, jute, silk, cotton and blends (and typically even synthetics) can be used as ground fabrics.

In crewel embroidery, stitch is used to hide tiny spans of house requiring a solid filling. Once operating plain stitch by itself, it is important that the house coated isn't overlarge, as a result of the stitches will loosen over time. The base materials utilized in crewel embroidery are durable and tightly woven, so that they will bear the load of the crewel yarn without sagging. The technique additionally needs a frame to secure the realm of the material being worked on, keeping the strain of the material tight in order that the stitches won't be distorted. Crewel needles are very sharp, as they have to be ready to penetrate dense, thick material. They even have elongated eyes that are supposed to facilitate threading, since thread colors in embroidery are typically switched, and craftspeople don't need to fight with their needles crewel embroidery is that the work of a proficient craftsperson, even once the bottom pattern is intended and applied by somebody else. The craftsman should be ready to produce sleek, clean stitches, employing a style of designs to feature texture and shading to the piece.

3. Applique Work

An appliqué is decorative needlework during which items of cloth ar sewed or stuck onto a bigger piece to create an image or pattern. this method is incredibly common in some styles of textile, however could also be applied to several materials. Within the context of ceramics, for instance, an appliqué could be a separate piece of clay extra to the first work, usually for the aim of decoration. The biggest variety of appliqué craftsmen are focused in Pipli, there are quite an few in Puri and extremely tiny numbers in Khallikote, Parlakhemundi and Boudh areas additionally.

The appliqué things are chiefly used throughout processions of the deities in their varied ritual outings. Things like Chhati, Tarasa and Chandua are used for the aim. The craft is historically experienced by a caste of skilled tailors, referred to as 'Darjis. The essential material for appliqué is textile. This lovely craft is sometimes practiced on dazzling red, purple, black, yellow, green and white material.

The craftsman initial prepares the bottom material within the form of sq., rectangle, and circle or oval that forms the background for the items of art. Appliqué motifs in different colors are then cut within the form of animals, birds, flowers, leaves, celestial bodies and geometric shapes then sewed onto the bottom textile in aesthetic arrangements. Nowadays, tiny mirrors and bright metal items are used to enhance its beauty. The traditional appliqué from Punjab is termed phulkari. In Andhra Pradesh, the blouses and headscarves worn by the Banjara tribal ladies don't seem to be solely decorated however additionally decorated with appliqué and mirror work. Rajasthan is additionally identified for its unique appliqué or gota, that is formed by stitching edges of zari ribbon onto material, to form elaborate patterns. Khandela in Shekhawati is best identified for manufacturing this stuff.

4. Mirror Embroidery

Mirror-Embroidery is a form of embroidery that attaches little pieces of mirrors mirror metal to fabric. particularly the dry desert sands and sunny areas of India, comprising of components of Gujarat, Rajasthan, Haryana and Delhi, sport strong colors and sheesha add most of their clothes, accessories and home furnishing product. In India, sheesha or mirror is on the market during a variety of shapes as well as spherical, square, triangular or two-dimensional figure. Mirror worked clothes, bags, bedspreads, wall hangings, cushion covers, etc are in nice demand all over the world. Specifically Gujarati embroidery is incorporating the sheesha or the mirror. It's used by the Jats of Banni. They cut the glass into totally different shapes and embroider it in the cloth. In Kathi embroidery the mirrors are used for eyes of birds and center of flowers. Mirror work is done on any cloth in sarees, blouse items, skirts etc. whether in home or workplace, party of picnic, in school or with colleagues – sheesha work clothes, home furnishing and accessories from India are a requirement for all people who like to look sensible and feel smart.

5. Kantha Embroidery

Kantha, a well-liked variety of embroidery that comes from west bengal, is a important image that displays the ability and talent of the rural ladies in Bengal. Kantha work involves advanced creative work done by the weavers with the mix of exquisite embroidery in decorative stitch. the normal work on the Kantha saris in the style of floral motifs, animals and birds figures and geometric shapes appearance surprisingly fabulous. nowadays this sort of embroidery is found on shawls, pillow covers, dupattas, and home furnishings as well.

There are seven differing kinds of Kantha. 1st is the 'Archilata Kantha' that involves the good mirror work with the wide, ornamental borders. The other is the 'Baiton Kantha', that are square wraps used for covering the books and valuables. the most noted is that the 'Lep Kantha' that ar rectangular wraps, used to cover the quilts. Then there's 'Oaar Kantha', found within the pillow covers in easy styles on which an ornamental border is really sewed later. Not less competitive is that the 'Sujani Kantha', the thick Kantha used as blankets on the special occasions. Last but not the smallest amount is the 'Rumal Kantha' that are used because the medicinal drug wipes or plate coverings. Renowned fashion designers in India acting on western designs and fashions, still build nice use of the ethnic traditions and their styles usually depict the made embroidery work of the sooner craftsmen, celebrating Indian textile tradition.

6. Phulkari

Phulkari embroidery is peculiar to Punjab. Phulkari literally means that flower craft. Baugh, which implies garden, is a Phulkari in which the whole surface is richly decorated by a continual connected style with skillful manipulation of the mending or stitch. Simple and sparsely decorated odini (head scarfs), dupatta and shawls, created for everyday use, are known as Phulkaris, whereas clothes that cover the whole body, created for special and ceremonial occasions like weddings and birth of a son, totally coated cloth is termed Baghs. Bright colors were invariably most popular and among these, golden yellow, red, crimson, orange, green, blue, pink etc, were the favored ones the sweetness of Phulkari depended a good deal on the color of the bottom material. Textile fabric that was hand spun and hand woven cotton material, was invariably used for embroidery.

The sew craft of Phulkari consists long short fixture stitches. It's a unique technique of embroidery in this it's worked entirely on the incorrect aspect of the material and also the pattern takes form on the proper aspect. There ar many varieties of Phulkari. The 'Chope' and 'Suber' were wedding Phulkar conferred to bride by her maternal relations throughout the wedding ceremony. The plain red / dark red khaddar scarf referred to as 'Saloo' was used for daily family wear. 'Til Patra' shawls have little embroidery and are inferior quality textile. 'Nilak' is worked on black or blue khaddar with yellow and crimson red pat. The Punjab Government's sales outlet, Phulkari, boasts of the most effective collection of this form of embroidery in the country.

BEFORE YOU BUY A SEWING MACHINE

If you are just a beginner in the sewing craft, then choosing the right sewing machine can be a confusing experience. There are so many machines available in the market with different features and prices that you will be at a loss wondering which to buy.

So, to make things easier, first of all decide what your needs are. You do not need to buy the most expensive machine with the most features. There is little point in investing in such a machine if you are not going to use most of the features given. Look for models that will fit your skill level and sewing needs. That is the most important consideration when you are shopping for a machine.

For example, think of what you are going to use the machine for. If you are interested in embroidery, you will need a machine that is different from someone who is only keen to use it for simple repair jobs. So, think ahead and plan what you are going to sew before buying your sewing machine.

With a list of features you need in hand, it will also be easier to narrow down your selection. Rather than being confronted with a wide range of choices, your needs will help weed out unsuitable models. That way, you can compare the features, prices and other details for the few remaining models before making your final decision.

To help you out, talk to experienced sewers for advice. They will be able to tell you which brand is more user-friendly and durable. You can also go through consumer reviews for the different machines. There are plenty of them on the sewing forums to give you an idea which model is a good buy for beginners and a value-for-money purchase.

On top of that, check out the various online stores for bargains. Sometimes, you may be able to get it cheaper online than through a retailer. However, an online purchase means you will need to learn how to set up the machine yourself and there is no sales staff to help you out should you encounter a problem. This is the disadvantage of buying a sewing machine through the web although you may be able to get it at a good price.

Lastly, sewing can be a very enjoyable hobby. Do not be afraid to take the plunge and get yourself a sewing machine. Using one if not as complicated as it seems with the right model. Moreover, with technological advancement, most machines for beginners now are very user-friendly that you can begin sewing almost immediately after purchase.

In its broadest sense, an appliqué is a smaller ornament or device applied to another surface. An appliqué is usually one piece. In the context of ceramics, for example, an appliqué is a separate piece of clay added to the primary work, generally for the purpose of decoration. The term is borrowed from French and, in this context, means "applied" or "thing that has been applied."

Appliqué was first discovered when clothes ripped and needed fixing so they used to sew over the top of the rip patches of different material otherwise known as patchwork. In the context of sewing, an appliqué refers to a needlework technique in which pieces of fabric, embroidery, or other materials are sewn onto another piece of fabric to create designs, patterns or pictures. It is particularly suitable for work which is to be seen from a distance, such as in banner-making. A famous example of appliqué is theHastings Embroidery.

Appliquéd cloth is an important art form in Benin, West Africa, particularly in the area around Abomey, where it has been a tradition since the 18th century and the kingdom of Danhomè.

Appliqué is used extensively in quilting. "Dresden Plate" and "Sunbonnet Sue" are two examples of traditional American quilt blocks that are constructed with both patchwork and appliqué. Baltimore album quilts, Broderie perse, Hawaiian quilts, Amish quilts and theralli quilts of India and Pakistan also use appliqué.

Having a creative hobby such as quilting, cross stitch or tapestry can be highly rewarding, but over time it can prove to be back breaking work. The process of creating an image or design takes a great many hours and the work requires a keen eye and persistent attention to detail. However problems can arise after long periods doing needlework as one can become so absorbed in their creation that they forget all about their posture. It's only when one stops for a break that they realise just how sore or strained their back has become. But don't despair because this can be rectified by using a kneeling chair.

Kneeling chairs have been popular since they were designed in 1979. Originally called the Balans Chair, the design has changed considerably over the years and the more modern designs feature lap-top stands, and for the keen needle worker there are quilting, tapestry and cross stitch stands. Chances are that currently you're using a standard tapestry frame which you sit at on a normal chair. This is not great for your posture as you tend to lean forwards towards your work, arching your spine unnaturally which can cause lower back pain.

When using a kneeling chair with an attached needlework frame such as a tapestry or quilting stand allows you to bring the work to you rather than the other way around. The sitting position one adopts on a kneeling chair is perfect for maintaining a good posture, and as the quilting or tapestry stand is fully adjustable the work can be placed in the perfect position; right under your nose if need be. By using a kneeing chair instead of a standard chair will vastly improve ones posture, especially when working on 'close-up' projects such as needlework or other handy crafts.

Kneeling chairs do have their share of bad press, with some people finding them uncomfortable after a short time. This problem arises from the relative dimensions of the chair not properly fitting the user. However by purchasing one which is tailored to your height means you should not experience any discomfort. Most manufacturers only produce kneeling chairs in a single 'standard' size. These are best avoided as the chair will either fit or it wont. Instead, look for a manufacturer that produces a variety of sizes so you can swap for the next size up or down should it not fit you perfectly.

Types

Applied pieces usually have their edges folded under, and are then attached by any of the following:

- Straight stitch, typically 2-3mm in from the edge.
- Satin stitch, all around, overlapping the edge. The patch may be glued or straight stitched on first to ensure positional stability and a neat edge.
- Reverse appliqué: several layers of material are stitched together, parts of the upper layers are cut away, and the edges are stitched down. The largest cuts are made in the topmost layer.

Appliqué is also used for school badges.

Appliqué and electronic sewing machines

Modern consumer embroidery machines quickly stitch appliqué designs by following a program. The programs have a minimum complexity of two thread colors, meaning the machine stops during stitching to allow the user to switch threads. First, the fabric that will be the background and the appliqué fabric are affixed into the machine's embroidery hoop.

The program is run and the machine makes a loose basting stitch over both layers of fabric. Next, the machine stops for a thread change, or other pre-programmed break. The user then cuts away the excess appliqué fabric from around the basting stitch. Following this, the machine continues on program, automatically sewing the satin stitches and any decorative stitching over the appliqué for best results.

Modern fashion

In this sense, appliqué refers to using fabric shapes or designs usually on the trim of a garment. This can be sewn or glued. Many appliques are more often imported from China and used here to keep costs down. This allows some nice pieces to be made at a lower price. Since many designers use appliques that are mass-produced, one can easily find matching accessories and such from competitive stores. Each may carry various items with the same applique.

Needle lace

Needle lace (also known as needlelace or needle-made lace) is a type of lacecreated using a needle and thread to stitch up hundreds of small stitches to form the lace itself. In its purest form the only equipment and materials used are a needle, thread and scissors. This form of lace making originated in Armenia where there is evidence of alace making tradition dating back to the pre-Christian era. Turkish needlelace is also very popular around the world. This form however arose separately from what is usually termed needlelace and is generally referred to as knotted lace. Such lace is very durable and will not unravel if one or more loops are broken.

Beginning in the 17th century in Italy, a variety of styles developed where the work is started by securing heavier guiding threads onto a stiff background (such as thick paper) with stitches that can later be removed. The work is then built up using a variety of stitches - the most basic being a variety of buttonhole or blanket stitch. When the entire area is covered with the stitching, the stay-stitches are released and the lace comes away from the paper. Needle lace is also used to create the fillings or insertions in cutwork.

Tapestry

Tapestry is a form of textile art, traditionally woven on a vertical loom, however it can also be woven on a floor loom as well. It is composed of two sets of interlaced threads, those running parallel to the length (called the warp) and those parallel to the width (called the weft); the warp threads are set up under tension on a loom, and the weft thread is passed back and forth across part or all of the warps.

Tapestry is weft-faced weaving, in which all the warp threads are hidden in the completed work, unlike cloth weaving where both the warp and the weft threads may be visible. In tapestry weaving, weft yarns are typically discontinuous; the artisan interlaces each coloured weft back and forth in its own small pattern area. It is a plain weft-faced weave having weft threads of different colours worked over portions of the warp to form the design. Most weavers use a naturally based warp thread such as linen or cotton. The weft threads are usually wool or cotton, but may include silk, gold, silver, or other alternatives.

The success of decorative tapestry can be partially explained by its portability (Le Corbusieronce called tapestries "nomadic murals"). Kings and noblemen could roll up and transport tapestries from one residence to another. In churches, they could be displayed on special occasions. Tapestries were also draped on the walls of castles for insulation during winter, as well as for decorative display.

In the Middle Ages and renaissance, a rich tapestry panel woven with symbolic emblems,mottoes, or coats of arms called a baldachin, canopy of state or cloth of state was hung behind and over a throne as a symbol of authority. The seat under such a canopy of state would normally be raised on a dais.

The iconography of most Western tapestries goes back to written sources, the Bible andOvid's Metamorphoses being two popular choices. Apart from the religious and mythologicalimages, hunting scenes are the subject of many tapestries produced for indoor decoration.

Historical development

Tapestries have been used since at least Hellenistic times. Samples of Greek tapestry have been found preserved in the desert of Tarim Basin dating from the 3rd century BC.

Tapestry reached a new stage in Europe in the early 14th century AD. The first wave of production originated in Germany and Switzerland. Over time, the craft expanded to France and the Netherlands. The basic tools have remained much the same since then.

In the 14th and 15th centuries, Arras, France was a thriving textile town. The industry specialised in fine wool tapestries which were sold to decorate palaces and castles all overEurope. Few of these tapestries survived the French Revolution as hundreds were burnt to recover the gold thread that was often woven into them. Arras is still used to refer to a rich tapestry no matter where it was woven.

By the 16th century, Flanders, the towns of Oudenaarde, Brussels, Geraardsbergen and Enghien had become the centres of European tapestry production. In the 17th century Flemish tapestries were arguably the most important productions, with many specimens of this era still extant, demonstrating the intricate detail of pattern and colour embodied in painterly compositions, often of monumental scale.

In the 19th century, William Morris resurrected the art of tapestry-making in the medieval style at Merton Abbey. Morris & Co. made successful series of tapestries for home and ecclesiatical uses, with figures based on cartoons by Edward Burne-Jones.

Kilims and Navajo rugs are also types of tapestry work. Tapestries are still made at the factory of Gobelins and a few other old European workshops, which also repair and restore old tapestries. The craft is also currently practiced by hobbyist weavers.

Jacquard tapestries, colour and the human eye

The term tapestry is also used to describe weft-faced textiles made on Jacquard looms. Before the 1990s tapestry upholstery fabrics and reproductions of the famous tapestries of the Middle Ages had been produced using Jacquard techniques but then artists such asChuck Close adapted the computerised Jacquard process to producing fine art.

Typically, tapestries are translated from the original design via a process resembling paint-by-numbers: a cartoon is divided into regions, each of which is assigned a solid colour based on a standard palette. However, in Jacquard weaving, the repeating series of multicoloured warp and weft threads can be used to create colours that are optically blended – i.e., the human eye apprehends the threads' combination of values as a single colour.

This method can be likened to pointillism, which originated from discoveries made in the tapestry medium. The style's emergence in the 19th century can be traced to the influence of Michel Eugène Chevreul, a French chemist responsible for developing the color wheel of primary and intermediary hues.

Chevreul worked as the director of the dye works at Les Gobelins tapestry works in Paris, where he noticed that the perceived colour of a particular thread was influenced by its surrounding threads, a phenomenon he called "simultaneous contrast." Chevreul's work was a continuation of theories of color elaborated by Leonardo da Vinci and Goethe; in turn, his work influenced painters including Eugène Delacroix and Georges-Pierre Seurat.

The principles articulated by Chevreul also apply to contemporary television and computer displays, which use tiny dots of red, green and blue (RGB) light to render colour, with each composite being called a pixel.

- The Trojan War tapestry referred to by Homer in Book III of the Iliad, where Iris disguises herself as Laodice and findsHelen "working at a great web of purple linen, on which she was embroidering the battles between Trojans andAchaeans, that Ares had made them fight for her sake." If such a tapestry had been made at the time, then it could explain how the battles were remembered in such great detail over the 400 or so years between the siege of Troy and the age of Homer.
- The Cloth of St Gereon - oldest European tapestry still extant.
- The Sampul tapestry, woollen wall hanging, 3rd-2nd century BC, Sampul, Ürümqi Xinjiang Museum.

- The Hestia Tapestry, 6th century, Egypt, Dumbarton Oaks Collection.
- The Bayeux Tapestry, which depicts the events surrounding the Battle of Hastings, is not a tapestry, but is instead embroidery. In June 2007, the tapestry was listed on UNESCO's Memory of the World Register.
- The Apocalypse Tapestry is the longest tapestry in the world, and depicts scenes from the Book of Revelation. It was woven between 1373 and 1382. Originally 140 m (459 ft), the surviving 100m are displayed in the Château d'Angers, in Angers.
- The six-part piece La Dame à la Licorne (The Lady and the Unicorn), stored in l'Hôtel de Cluny, Paris.
- The Devonshire Hunting Tapestries, four Flemish tapestries dating from the mid-fifteenth century depict men and women in fashionable dress of the early fifteenth century hunting in a forest. The tapestries formerly belonged to the Duke of Devonshire and are now in theVictoria and Albert Museum.
- The Hunt of the Unicorn is a seven piece tapestry from 1495 to 1505, currently displayed at The Cloisters, Metropolitan Museum of Art in New York.
- The tapestries for the Sistine Chapel, designed by Raphael in 1515-16, for which theRaphael Cartoons, or painted designs, also survive.
- The Wawel Tapestries, (mid 16th century) a collection of 134 tapestries at the Wawel Castle in Kraków, Poland displaying various religious, natural, and royal themes. These famous tapestries, created in Arras, were collected by Polish Kings Sigismund I the Oldand Sigismund II Augustus.
- The Valois Tapestries are a cycle of 8 hangings depicting royal festivities in France in the 1560s and 1570s
- The New World Tapestry is a 267 feet long tapestry which depicts the colonisation of the Americas between 1583 and 1648, displayed at the British Empire and Commonwealth Museum; this is not (strictly speaking) a tapestry, but is instead embro-idery.
- The biggest collection of Flanders tapestry is in the Spanish royal collection, there is 8000 metres of historical tapestry from Flanders, as well as Spanish tapestries designed by Goya and others. There is a special museum in the Royal Palace of La Granja de San Ildefonso, and others are displayed in various historic buildings.

- Pastoral Amusements, also known as "Les Amusements Champêtres", a series of 8 Beauvais Tapestries designed by Jean Baptiste Oudry between 1720 and 1730.
- The Prestonpans Tapestry is a 104 metres long embroidery which tells the story of Bonnie Prince Charlie and the Battle of Prestonpans.
- Christ in Glory, (1962) for Coventry Cathedral designed by Graham Sutherland. Up until the 1990s this was the world's largest vertical tapestry.
- The Quaker Tapestry (1981–1989) is a modern set of embroidery panels that tell the story of Quakerism from the 17th century to the present day.

Woolen or woollen is a type of yarn made from carded wool. Woolen yarn is soft, light, stretchy, and full of air. It is thus a good insulator, and makes a good knitting yarn. Woolen yarn is in contrast to worsted yarn, in which the fibers are combed to lie parallel rather than carded, producing a hard, strong yarn.

The woolen and worsted process both require that the wool (and other similar animal fibers, cashmere, camel, etc.) are cleaned before mechanical processing. Woolen and worsted nomenclatures apply only to the textile processing of animal fibers, but it has become common to include fiber blends under these terms.

The resultant fabrics will be classified as being either woolen or worsted, but this designation is assigned during fiber processing and yarn formation, not in the cloth or finished garment.

A woven woolen fabric is one which is subjected to fabric finishing techniques designed to add a directional pile - in that the end consumer can 'stroke' the garment in a single direction (shoulder to cuff etc.), such as a casual jacket. This feels like the fibers are directionally arranged.

Woolen yarn formation is also very common for knitwear, where the resultant garment has some bulk and the requirement for visual aesthetics (of fiber alignment) is minimal. The worsted processing route is more complex and requires the removal of short fibers and the use of a focused mechanical process to make the individual fibers parallel to each other. The yarn formation process is significantly more comprehensive and results in a very sleek yarn which will offer a clean looking woven fabric, such as for suitings. The worsted process is significantly more expensive and is seldom used for knitwear.

HANDSPINNING WOOLEN YARN

Woolen yarn is handspun using the long draw technique, and the yarn is spun from a rolag. Most handspinners make a blend of a woolen and worsted yarn, using techniques from both categories, and thus ending up with a mix. The first step to spin a true woolen yarn, however, is to card the fiber into a rolag using handcarders. The rolag is spun without much stretching of the fibers from the cylindrical configuration.

This is done by allowing twist into a short section of the rolag, and then pulling back, without letting the rolag change position in your hands, until the yarn is the desired thickness.

The twist will concentrate in the thinnest part of the roving, thus when the yarn is pulled, the thicker sections with less twist will tend to thin out. Once the yarn is the desired thickness, enough twist is added to make the yarn strong. Then the yarn is wound onto the bobbin, and the process starts again.

Other meanings

"Woolen", as an adjective, also refers to a thing which is composed of wool, in the same vein as wheaten, golden, and the somewhat rarer silvern and leathern.

Thread (yarn)

A quilter's thread options are truly mind boggling. Thousands of color choices are just the beginning, and possibly the easiest aspect of choosing thread. Beyond that, you'll find threads specifically for quilting, and discover that hand quilting threads have characteristics that aren't a plus for machine quilting tasks.

Threads developed for machine embroidery and other decorative techniques are varied, and threads that work best for piecing are often not labeled for that purpose, making selection difficult. I hope this quick thread guide will help you wade through the maze of threads quilters use most often.

A Quick Look at Cotton and Polyester Thread Creation

Cotton thread is created by spinning cotton fibers together, and then pulling and twisting a narrow strand of yarn away from the mass. Individual strands of yarn, each called a ply, can be twisted together to create a stronger thread.

Polyester, a synthetic product, can be spun together in a similar way to create threads that look like cotton, but have more stretch. Polyester can also be drawn out into long, continuous-filament threads. All-polyester thread may, over time, cut cotton fabric.

Another type of thread is made with a polyester core encased in cotton, resulting in a slightly stretchy thread, but with a traditional look and feel. Most quilters prefer all-cotton threads over the cotton/poly combo. Spool of all purpose sewing thread, close-up shows texture of 2-ply Z-twist mercerized cotton with polyester core. Thread is a kind of yarn used for sewing.

Decorative Rayon, Nylon and Metallic Threads

Rayon is derived from cellulose, but is not classified as a natural fiber because the transformation requires quite a bit of manipulation. Colorful rayon threads are very popular with quilters, and are typically used for machine embroidery and other decorative work.

Nylon is a synthetic product used to make transparent monofilament thread, which becomes fairly invisible when used for machine quilting. However, it can melt under an iron, discolors, and becomes brittle as it ages. A fine, transparent polyester thread is a more durable choice.

Metallics typically have a core of nylon or polyester that's covered with a decorative product. Quality metallic threads also have an outer coating to help protect the delicate metallic layer.

Additional Threads

You'll encounter threads made from other natural materials, including wool and silk.

- Wool threads are typically thicker than other threads, and sometimes used to embellish a Folk Art or homespun-look project.
- Silks are often used for applique — they are fine and make stitches seem to disappear. Silk threads are also a good choice for beading on fabric.

Water soluble threads dissolve when a project is washed. They are used for basting, or for any task where temporary stitches are handy.

Fusible threads are used to sew a typical seam, but when pressed they fuse the sewn fabrics together. Binding and applique are two potential uses for fusible threads.

You'll find other specialty threads when you explore thread manufacturer web sites.

Wading Through Thread Sizes

Unfortunately, there is no single system used to describe thread sizes.

One common size designation is depicted as a fraction, such as 50/3. The first number reveals the thread's weight and the second tells us the number of plies — a 50/3 thread has a weight of 50 and is made with 3 plies of yarn. With this system, thread weight decreases as weight numbers increase. A 50/3 thread is commonly used for piecing, but other equivalent sizes work just as well.

Another sizing system indicates only thread weight.

Sizing systems are complex, and often inaccurate as threads that are imported and exported around the world are re-labeled for a new country. Choose threads based on recommendations from manufacturers and experienced quilt makers.

More Thread Terminology

Shopping for thread may introduce you to new terms. A few words you might hear are:

Mercerized: Cotton thread is processed with chemicals that give it more luster, improve strength and help it be more susceptible to dyes. The process makes thread more fuzzy, which is reduced by either a gassing or singing process.

Crocking: Crocking occurs when dye on the surface of thread (or fabric) rubs off onto other materials.

Denier: Denier is a sizing method that's often used for continuous filament threads. The number indicates the weight in grams of 9,000 meters of the thread.

Threads for Piecing

All-cotton threads are readily available, and are the best choice for piecing our quilting cottons. You can help avoid future wear at the seams by choosing a thread that's no stronger than the fabric, so avoid polyester threads and overly strong cotton threads. A hefty thread can also affect seam allowances and accuracy, by taking up too much space in the seam and by causing excess bunching.

The threads below are all good choices for piecing, but so are many others. Use my list as a starting point, but be sure to ask staff at your local quilt shops for their recommendations.

Aurifil An almost lint-free 50 weight thread.

YLI Select A 40/2 thread made from Egyptian cotton.

Gutermann Cotton A 50 weight cotton thread.

Sewing with Decorative Threads

Decorative threads (for hand and machine embroidery and for machine quilting) include the rayons and metallics mentioned above. One way to preview these threads is to browse manufacturer web sites. Take some time to explore the sites — most have educational resources to help you choose threads, and they often suggest appropriate needles and other supplies that are required for specific jobs.

Quilting Threads

Quilters use all sorts of threads for quilting, from cottons to decoratives. But remember one thing — threads developed specifically for hand quilting should not be used in your sewing machine, because they are coated with a glaze that is not machine-compatible. Beyond that, the sky's the limit.

Manufacturer Web sites are once again a good source of information for threads suitable to hand and machine quilting. Do keep in mind that you must often change your sewing machine's tension when you machine quilt, use a different thread in the bobbin, and choose hand and machine needles that will deliver the thread intact, without damage to its integrity.

Bottom Line

This chapter helps new quilters begin to gain a better understanding of the threads quilters can use in their projects, but it's simply a basic overview. Entire books can be written about threads and their uses, and the selections are constantly expanding. Ask questions about threads — your local guild and quilt shops are excellent resources. Don't be afraid to experiment with threads, because experimentation is such a good way to learn. Once you have sewn with several different threads, you'll have a much better feel for future selections.

Materials

Thread is made from just about everything. Following table lists common materials, a general description and what they are supposed to be good for. If your machine will sew with the thread, any thread can used for just about any purpose. This is very useful for someone who is trying to learn sewing. Please improve/ expand this table.

DIFFERENT WEIGHTS

Weight

Weight is a way of indicating the thickness of a thread. The most common weight system specifies the length of the thread in kilometers required to weigh one kilogram. Therefore, a greater weight number indicates a thinner thread. American standard of thread weight (wt) was adopted from Gunze Count standard (Japan). Gunze Count standard uses two numbers separated by a forward slash. The first number indicates the weight of the internal threads and the second number indicates the number of threads wrapped together to make the finished thread.

It is common to wrap three strands of the same weight to make one thread (as indicated in the chart), any number of strands may be used and sometimes only two strands are wrapped together to made the finished thread. The American standard drops the specification of the number of strands which are twisted together. Therefore it is does not accurately describe thread.

Thread Weight Table

	Weight	Gunze Count	Common Use
Light	60 wt	#60/3	bobbin or appliqué
Thin	50 wt	#50/3	bobbin or appliqué
Regular	40 wt	#40/3	Quilting
Upholstery	30 wt	#30/3	Decorative
Heavy	20 wt	#20/3	Decorative

Denier

A denier weight specification states how many grams 9,000 meters of the thread weighs. Unlike the common thread weight system, the greater the denier number, the thicker the thread. The denier weight system, like the common weight system, also specifies the number of strands of the specified weight which were wrapped together to make the finished thread.

Tex

Tex is the mass in grams of 1000 meters of thread. If 1,000 meters weighs 25 grams, it is a tex 25. Larger tex numbers are heavier threads. Tex is used more commonly in Europe and Canada.

Conversion Information

THREAD WEIGHT CONVERSION TABLE

Converting From	Converting T·	Method
Weight	Denier	9000 / weight
Weight	Tex	1000 / weight
Denier	Weight	9000 / denier

Denier	Tex	denier x 0.111
Tex	Denier	tex x 9
Tex	Weight	1000 / tex
For Example: 40 weight = 225 denier = Tex 25		

Textile yarns are measured in various units, such as: the denier and tex (linear mass density of fibres), super S (fineness of wool fiber), worsted count, woolen count, cotton count (or Number English Ne), Number metric (Nm) and yield (the inverse of denier and tex). Yarn is spun thread used for knitting, weaving, or sewing. Thread is a long, thin strand of cotton, nylon, or other fibers used in sewing or weaving.

Both yarn and thread are measured in terms of cotton count and yarn density. Fabric is cloth, typically produced by weaving or knitting textile fibers, and is measured in units such as mommes (momme is a number that equals the weight in pounds of a piece of silk if it were sized 45 inches by 100 yards), thread count (a measure of the coarseness or fineness of fabric), ends per inch (e.p.i), and picks per inch.

Denier or den is a unit of measure for the linear mass density of fibers. It is defined as the mass in grams per 9,000 meters. In the International System of Units the text is used instead. The denier is based on a natural standard: a single strand of silk is approximately one denier. A 9,000-meter strand of silk weighs about one gram. The term denier is from the French denier, a coin of small value (worth 1D 12 of a sou). Applied to yarn, a denier was held to be equal in weight to 1D 24 of an ounce. The term microdenier is used to describe filaments that weigh less than one gram per 9,000 meter length.

One can distinguish between filament and total measurements in deniers. Both are defined as above but the first only relates to a single filament of fiber (also commonly known as denier per filament or DPF), whereas the second relates to a yarn, an agglomeration of filaments.

The following relationship applies to straight, uniform filaments:

DPF = total denier / quantity of uniform filaments

The denier system of measurement is used on two- and single-filament fibers. Some common calculations are as follows:

1 denier = 1 gram per 9 000 meters

= 0.05 grams per 450 meters (1D 20 of above)

= 0.111 milligrams per meter

In practice measuring 9,000 meters is both time-consuming and wasteful. Usually a sample of 900 meters is weighed and the result multiplied by 10 to obtain the denier weight.

- A fiber is generally considered a microfiber if it is 1 denier or less.
- A 1-denier polyester fiber has a diameter of about 10 micrometers.
- Denier is used as the measure of density of weave in tights and pantyhose, which defines their opacity.

One can calculate the diameter of a filament given its weight in denier with the following formula:

$$\varnothing = \sqrt{\frac{4.444 \times 10^{-6} \cdot \text{denier}}{\pi \rho}}$$

where ρ represents the material's density in grams per cubic centimeter and the diameter is in cm.

Tex

Tex is a unit of measure for the linear mass density of fibers and is defined as the mass in grams per 1000 meters. Tex is more likely to be used in Canada andContinental Europe, while denier remains more common in the United States and United Kingdom. The unit code is "tex". The most commonly used unit is actually the decitex, abbreviated dtex, which is the mass in grams per 10,000 meters. When measuring objects that consist of multiple fibers the term "filament tex" is sometimes used, referring to the mass in grams per 1000 meters of a single filament.

Tex is used for measuring fiber size in many products, including cigarette filters, optical cable, yarn, and fabric.

One can calculate the diameter of a filament given its weight in dtex with the following formula:

$$\varnothing = \sqrt{\frac{4 \times 10^{-6} \cdot \text{dtex}}{\pi \rho}}$$

where represents the material's density in grams per cubic centimeter and the diameter is in cm.

Tex (g/km)	Yield (yards/#)
550	900
735	675

1100	450
1200	413
2000	250
2200	225
2400	207
4400	113

S or super S number

Super S or S number is an indirect measure of the fineness of the wool fiber. It is most commonly seen as a label on wool suits and other tailored wool apparel to indicate the fineness of the wool fiber used in the making of the apparel. The numbers may also be found on wool fabric and yarn.

Worsted count

Yarn spinning factory

Worsted count (or spinning count) is an indirect measure of the fineness of the fiber in a worsted wool yarn expressed as the number of 560-yard (1 yard = 0.9144 meters) lengths (hanks) of worsted yarn that a pound (0.45359237 kilograms) of wool yields. The finer the wool, the more yarn and the higher the count. It has been largely replaced by direct measures.

Yield

Similar to tex and denier, yield is a term that helps describe the linear density of a roving of fibers. However, unlike tex and denier, yield is the inverse of linear density and is usually expressed in yards/lb.

YARN AND THREAD

Cotton Count

- Cotton Count is another measure of linear density. It is the amount of skein material measured in hanks (840 yards) needed to create one pound. Under this system, the higher the number, the finer the yarn. In the United States a cotton count between one and 20 are referred to as coarse counts. A regular single knit T-Shirt can be between 20 and 40 count, fine bed sheets are usually in the range of 40 to 80 count. The number is now widely used in the staple fiber industry.
- Hank: A length of 7 leas or 840 yards

One Lea - 120 yards

Yarn Length

Yarn Length in meters = Yarn count × 1693 × Yarn Weight in kg

English Cotton count (Nec)is an indirect counting system i.e. the higher the number the finer the yarn.

- Thread: A length of 54 in (the circumference of a warp beam)
- Bundle: Usually 10 lbs
- Lea: A length of 80 threads or 120 yards
- Denier: this is an alternative method. It is defined as a number that is equivalent to the weight in grams of 9000 m of a single yarn. 15 denier is finer than 30 denier.
- Tex: is the weight in grams of 1 km of yarn.

To convert Denier to Cotton Count *5315 - from Tex * 590.5 Tex is 1/9 Denier

Thread

Thread is a cotton yarn measure, equal to 54 inches.

Fabric

Mommes

Mommes (mm) are units of weight traditionally used to measure the surface density of silk. It is akin to the use of thread count to measure the quality of cotton fabrics, but is calculated in a very different manner. Instead of counting threads, the weight in mommes is a number that equals the weight in pounds of a piece of silkif it were sized 45 inches by 100 yards. This is because the standard width of silk is 45 inches wide, though silk is regularly produced in 55-inch widths, and, uncommonly, in even larger widths.

Silk can also be measured by weight in grams. 1 momme = 4.340 grams per square meter; 8 mommes is close to 1 ounce per square yard or 35 grams per square meter.

The usual range of momme weight for different weaves of silk are:

- Habutai - 5 to 16 mm
- Chiffon - 6 to 8 mm (can be made in double thickness, i.e. 12 to 16 mm)
- Crepe de Chine - 12 to 16 mm
- Gauze - 3 to 5 mm
- Raw silk - 35 to 40 mm (heavier silks appear more 'wooly')
- Organza - 4 to 6 mm
- Charmeuse - 12 to 30 mm

The higher the weight in mommes, the more durable the weave, and the more suitable it is for heavy-duty use. And, the heavier the silk, the more opaque it becomes. This can vary even between the same kind of silk. For example, lightweight charmeuse is translucent when used in clothing, but 30-momme charmeuse is opaque.

Thread count

Thread count is a measure of the coarseness or fineness of fabric. It is measured by counting the number of threads contained in one square inch of fabric or one square centimeter, including both the length (warp) and width (weft) threads. The thread count is the number of threads counted along two sides (up and across) of the square inch, added together. It is used especially in regard to cotton linens such as bed sheets, and has been known to be used in the classification of towels.

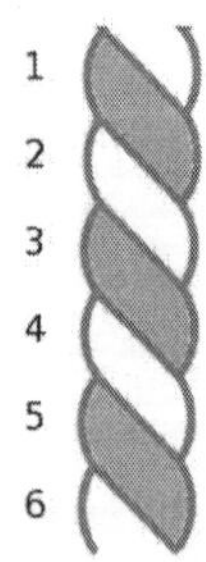

Image showing how to determine the number of twists per inch in a piece of yarn

Thread count is often used as a measure of fabric quality, so that "standard" cotton thread counts are around 150 while good-quality sheets start at 180 and a count of 200 or higher is considered percale. Some, but not all, of the extremely high thread counts (typically over 500) tend to be misleading as they usually count the individual threads in 'plied' yarns (a yarn that is made by twisting together multiple finer threads).

For marketing purposes, a fabric with 250 two-ply yarns in both the vertical and horizontal direction could have the component threads counted to a 1000 thread count although "according to the National Textile Association (NTA), which cites the international standards group ASTM, accepted industry practice is to count each thread as one, even threads spun as two- or three-ply yarn.

The Federal Trade Commission in an August 2005 letter to NTA agreed that consumers 'could be deceived or misled' by inflated thread counts. In 2002, ASTM proposed a definition for "thread count" that has been called "the industry's first formal definition for thread count". A minority on the ASTM committee argued for the higher yarn count number obtained by counting each single yarn in a plied yarn and cited as authority the provision relating to woven fabric in the Harmonized Tariff Schedule of the United States, which states each ply should be counted as one using the "average yarn number."

Ends per inch

Ends per inch (or e.p.i.) is the number of warp threads per inch of woven fabric. In general, the higher the ends per inch, the finer the fabric is.

Ends per inch is very commonly used by weavers who must use the number of ends per inch in order to pick the right reed to weave with.

The number of ends per inch varies on the pattern to be woven and the thickness of the thread. Plain weaves generally use half the number of wraps per inch for the number of ends per inch, whereas denser weaves like a twill weave will use a higher ratio like two thirds of the number of wraps per inch. Finer threads require more threads per inch than thick ones, and thus result in a higher number of ends per inch.

The number of ends per inch in a piece of woven cloth varies depending on what stage the cloth is at. Before the cloth is woven the warp has a certain number of ends per inch, which is directly related to what size reed is being used. After weaving the number of ends per inch will increase, and it will increase again after being washed.

This increase in the number of ends per inch (and picks per inch) and shrinkage in the size of the fabric is known as the take-up. The take-up is dependent on many factors, including the material and how tightly the cloth is woven. Tightly woven fabric shrinks more (and thus the number of ends per inch increases more) than loosely woven fabric, as do more elastic yarns and fibers.

Picks per inch

Picks per inch (or p.p.i.) is the number of weft threads per inch of woven fabric. A pick is a single weft thread, hence the term. In general, the higher the picks per inch, the finer the fabric is.

5

UNDERSTANDING YARN COUNT

Yarn count is a numerical value that indicates the coarseness or fineness of a yarn. Learn about the definition and formula of yarn count and get to know the direct and indirect systems of calculating yarn count. Some people try to count sheep when they go to bed. For others, however, a good night's sleep only comes from thinking about yarn count. Before any textile, including bed sheets, can be made, the manufacturers have to produce the yarn. The quality of a sheet depends on its yarn count, which is a numerical value that tells you how fine or coarse a yarn is. In technical terms, it's the value of the linear density (the diameter) to which that yarn was spun. When helping you fall asleep, this is the count that will matter.

Yarn is a long continuous length of interlocked fibres, suitable for use in the production of textiles, sewing, crocheting, knitting, weaving, embroidery and ropemaking. Thread is a type of yarn intended for sewing by hand or machine. Modern manufactured sewing threads may be finished with wax or other lubricants to withstand the stresses involved in sewing. Embroidery threads are yarns specifically designed for hand or machine embroidery.

Since yarn is the compressible textile material it is not possible to measure its fineness by any of the testing instrument so it is usually expressed in terms of the linear density or count. In short the count of yarn is a numerical expression which defines its fineness or size of the yarn. It also expresses weather the yarn is thick or thin. According to the textile institute yarn count is a number of indicating the length per unit weight of the yarn or weight per unit length.

YARN COUNT TESTING METHOD

Wrap reel method

Skein gauge method

Drying oven method

Analytic balance method

What is yarn count numbering?

Yarn numbers is measures of the coarseness or fineness of a yarn indicating the length per unit mass or mass per unit length and the method of calculation of count number is called as yarn count numbering system.

Importance of yarn count numbering system:

The fineness of yarn cannot be expressed easily in terms of diameter as in the case of wires and cords we measure its diameter by using instrument such as micrometer or calipers. But in case of yarn we cannot measure its diameter by using micrometer and caliper because most of the yarns are relatively soft and compressible, also the cross section of the spun yarn is not fully circular and it contains variations in thickness because presence of thick and thin places in the yarn and the continuous filament yarns also suffer from this drawback. Therefore yarn numbering systems like special methods are used to measure the fineness and coarseness of yarn.

Types of yarn numbering system:

Direct yarn numbering system

Indirect yarn numbering system

Linear Density or Count:

Yarn count is a numerical value which expresses the coarseness or fineness (diameter) of the yarn and also indicates the relationship between length and weight (the mass per unit length or the length per unit mass) of that yarn. The fineness of the yarn is usually expressed in terms of its linear density or count. If you see the below figure, concept of yarn count will be clear.

According to Textile Institute, "Count is a number indicating the mass per unit length or the length per mass unit of a yarn".

CALCULATING YARN COUNT

The yarn count can tell you a lot about a yarn's durability, strength, and comfort. It's an important number, and that means it's one that you don't want to estimate.

You can probably guess where this is going. We're going to have to talk about math and in a textiles course, no less.

In its most basic sense, the yarn count represents either the mass per unit length, or length per unit mass of the spun yarn. Let's look at two different ways to calculate this number, both of which can ultimately tell us how fine or coarse the yarn is but involve different routes of getting there.

Types of Yarn Counting System

There are two types of yarn counting system:

Direct system

Indirect system

Direct System

Let's start with the first method for calculating yarn count: the direct system, which calculates the weight of a yarn by treating the length of the yarn as the constant in the formula. Basically, it helps you answer the question, 'How much does a consistent length of yarn weigh?' That's what the yarn count represents. In this system, a higher yarn count represents a heavier and, therefore, coarser yarn.

The basic formula for the direct system looks like this:

N = (W/l) / (L/w)

In this formula, N is the Yarn Count, W is the weight of a sample of yarn, l is the unit of length, L is the length of the sample, and w is the unit of weight. W and L come from the specific yarn being used, but what about w and l? These numbers come from the metric we're using to calculate yarn count. Here are three units we often use in the direct system:

Tex: Grams per 1000m (1 km) of yarn

Denier: Grams per 9000m (9 km) of yarn

Pounds per Spindle: Pounds per 14,400 yards of yarn

For example, a Tex would be the weight in grams in one kilometer of yarn. In our formula for the direct system, the unit of length would be one kilometer, and the unit of weight would be a gram. From there, all we need is the actual weight and length of the sample, and we can calculate the yarn count.

INDIRECT SYSTEM

Count is measured as the length of yarn per unit weight. The common feature of all indirect count systems is the weight of yarn fixed, and the length of yarn varies according to its fineness. Less heavy yarn has greater length per unit weight. Indirect system is used in the English system.

In this yarn counting system, If increase Count then decrease Dia.

Indirect yarn count system:

Numbering System	Unit of Length (l)	Unit of Weight (w)
English cotton count, N_e	840 yards (yds)	1 pound (lb)
Metric count, N_m	1000 metres / 1 km	1 kg
Woollen count (YSW)	256 yards	1 pound (lb)
Woollen count (Dewsbury)	1 yard	1 ounce (oz)
Worsted count, NeK	560 yards	1 pound (lb)
Linen count, NeL	300 yards	1 pound (lb)

Types of yarn count:

Normal count

Conditional count

Average count

Resultant count

Relation between yarn diameter and count:

Assuming an apparent specific volume for cotton yarn is 1.1:

Specific volume of yarn is 1.1 cc/gm

The mass of 1.1 cc yarn = 1 gm

If the yarn count is N tex then, we get

The length of N gm yarn is 1000m

The length of 1 gm yarn is 1000/N m=100000/N m

So that, length of 1.1 cc yarn is 100000/N m

We know that,

Cross sectional area × Length = Volume

Measurement of Yarn Counting System:

Wrap reel and balance method

Beesley's yarn balance

Knowles balance

Quadrant balance method

Sliver, roving count by measuring drum

Count data system

Auto sorter by Uster

Yarn Count Selection in Knitting:

Knitted fabric is made with the help of yarn loops. Yarn of different counts is used to produce fabric of different grammage. There is also a need to calculate optimum production of knitting machines. It is the job of knitting manager to do certain calculation for proper use of machines and production of fabric according to the demands of the customer. In this article I will discuss for selection of yarn count in knitting.

MOST SUITABLE COUNT FOR KNITTING MACHINES

As we know needle hook has to take yarn to convert it into a loop and finally latch has to close the needle hook so that loop is properly held by the needle hook and ultimately this helps in passing new loop through the previously held loop. It is clear from this explanation that there should be a proper balance between needle hook size and the thickness of the yarn or filament. If the yarn is thicker than needle hook then there will a chance that needle hook will not able to hold this loop and consequently there will be a small hole in the fabric. If the situation is reverse, means yarn is thinner than the size of the needle hook then the fabric produced will look like a net.

Both situations are not wanted. This situation demands a balance between needle hook size and count of yarn. It is worth to note that needle hook size depends upon the machine guage. Furthermore for different garments, fabric of different grammage is required. Every time knitter has to decide about the yarn count. There are many ways for the selection of proper count. In the following lines we will discuss most common methods to select count for different machines of different guage. It is also important to note that selection of yarn counts also depends upon the machine manufactures and type of machines, like, single and double knit machine. However a general guideline will be given hereunder.

As a thumb rule knitting experts prefer to use such knitting machine whose gauges is near to count of yarn (English count) i.e. for 20-gauge machines most suitable yarn count is 20s. This rule is has certain limitations, like, for 28-gauge yarn of 26s to 30s is most suitable. But for very fine counts this rule is not applicable and also machines have maximum gauge 32. Normally fine counts are not used as such rather they are make double, like count 60s double, which means that net count is near to 30s. And this 60 double count is suitable for 30-gauge machine.

To solve this problem some authors have suggested following formulas:

For single knitting machine,

Suitable count = G*G/18

For Double knitting machine,

Suitable count= G*G/8.4

Where G is gauge of knitting machine

Some knitting machine manufacturers suggest a range of yarn count for their machine. There is another way to solve this problem and that is to take help from old record. Every firm is producing many types of fabrics and on the basis of experience they develop a database for ready reference. In the following line we give a table for guidance (table is under construction). One can get a ready reference from the table to produce fabric of certain grammage. We are also giving expected width of fabric after wet processing. This table can provide just a reference. Knitters have to decide by themselves after doing a trial production, since there are many more factors, which can affect yarn and gauge selection process.

KNITTING MACHINE PARAMETERS

Every knitting machine is made to fulfill certain demands of the customer. There are number of characteristics of machine which are intimated by the machine manufacturers while delivering the machine to customers/users. It is helpful for the user to be well aware about these parameters. Furthermore machine specifications are given in different unit. We will explain these parameters and will also give the conversion factors to convert parameters from one system to other.

Machine Gauge:

As per Oxford Dictionary the word "gauge" is a noun and as well as verb. It is used to measure level of anything or for an instrument to measure width, length or height of anything. In knitting it is used to express the number of needle in a unit length of the needle bed. This needle bed may flat or circular. In double knit circular machine it is used for cylinder and as well as dial. Generally gauge is defined as number of needles per inch. According to German standard DIN 60917 (Iyer et al1995) alphabet "E" is used to denote knitting machine gauge.

Machine Pitch:

As per German DIN 62125 (Iyer et al1995) the notation "gauge" is to be avoided in the future. Rather they prefer to use notation "pitch" for comparison purpose. Machine Pitch means the distance between the centers of two neighboring needles. It is denoted with small "t". It is given in mm.

Theory:

Yarn count: The yarn count is a numerical expression which defines its fineness or coarseness. It also expresses weather the yarn is thick or thin. A definition is given by the textile institute – "Count is a number which indicates the mass per unit length or the length per unit mass of yarn."

To determine the thread count or yarn count we use five instruments with wrap reel & analytical balance. Beesley balance is one of them. It is working on the principle of fixed weight and fixed length system

Objectives of Experiment:

To know about thread count or yarn count.

To know about template & beesley balance.

To know how to determine the thread count by template & beesley balance.

Required Instrument:

Template

Beesleys Balance

Knife

Marking pen

Needle

Template:

Template is a length measuring instrument which contains eight arms. When a sample has short length (Fabric) which we cannot determine by the wrap reel then we use template by using that we can determine the length.

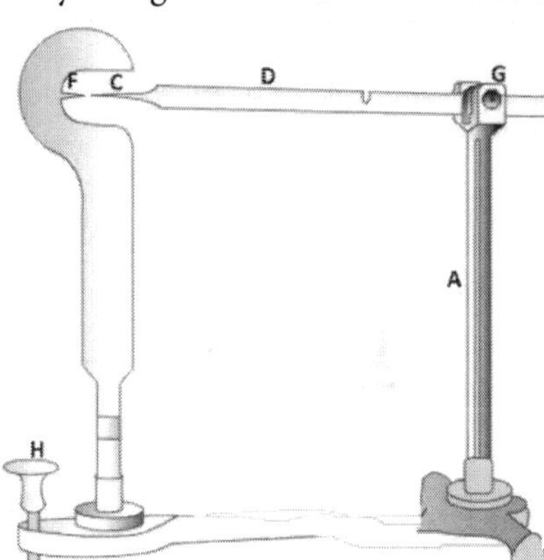

Figure:Beesley Balance

A-Pillar; B-Beam; C-Pointer; D-Notch; E- Sample hook; F-Detam Line; G-Pivot; H-Levelling Screw

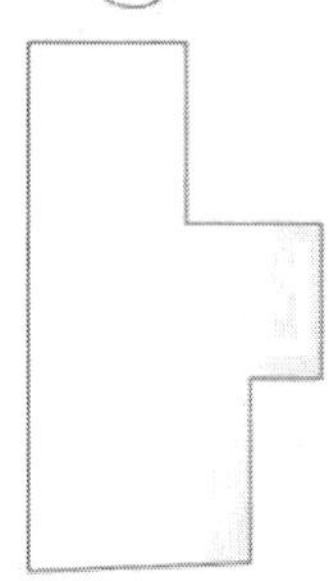

Figure: Templet

The following length is used to determine the count:

Cotton: 4.32 Inch

½ cotton: 2.16 Inch

Linen: 1.543 Inch

Wool: 1.315 Inch

Worsted: 2.88 Inch

Beesleys Balance:

Beesleys yarn balance consists of a pillar A which carries a cross beam B, fulcrumed at knife edge at the point G (See figure in above). At one end the cross beam is a hook E upon which the yarn to be tested can be placed. The other end of the beam tappers to a point C. When the beam is in balance, the pointer will coincide with the detum line. The pillar is mounted on the base. The whole instrument is leveled by a leveled screw at one end of the base. The cross has a small notch at the point D to take the counter weight or rider.

DATA COLLECTION

Sample No	Amount of yarn	Yarn Count	Average Count
01	29	29	30s
02	30	30	
03	31	31	
04	30	30	
05	30	30	
06	29	29	
07	30	30	
08	31	31	
09	30	30	
10	30	30	

Result: The cotton sample fabric warp thread count is 30s

Working Procedure:

Collect sample by appropriate sampling method.

This Sample conditioning at testing atmosphere.

Marked the fabric by using template

Cut that's fabric by knife according to the marking.

The pointer is set directly opposite to the detum line, with no material and counter weight in their proper places, by adjusting the leveling screw. The counter weight for the particular length which is supplied with the instrument is chosen and suspended at the notch D. (For full cotton the large rider is placed in the notch and for ½ cotton small rider is placed)

Now yarn is withdrawn from sample and placed sample hook until the pointer comes in level with the detum line.

At that stage the threads are taken out and counted which givers directly the cont of yarn teken for testing.

There 30 threads in the sample hook at the balanced condition so the count of the yarn is 30s

Precaution:

We should yarn withdrawn from fabric very carefully.

Marked fabric very carefully.

Conclusion:

It is the very easy process but if we cannot use template properly according to thread type and idle weight than the result will be faulty.

Importance of yarn count:

The main purpose of manufacturing of the yarn is to convert it in to the fabric form. Usually 90% of the yarn is converted into the fabric form for making its use and the remaining is used as it is in the yarn form. If the yarn is used as it is then there is no so much need of measuring its dimensions, but if it is to be converted in to the fabric form then in order to manufacture the required fabric, the yarn parameters plays vital role. The fabric parameters are decided from the use of that fabric for which it is to be used, and those fabric parameters are inculcated from the yarn parameters.

The fabric construction is varied with respect to its end use. The main important fabric factors deciding its construction are, yarn spacing and yarn thickness. Out of this the yarn spacing can be measured by counting the number of yarns per inch with use of magnifying glass. The yarn thickness i.e. yarn diameter (assuming that the yarn is almost circular in cross section) is not an easy process even by any of the means due to the following problems.

6

'KNITTING' GROWTH THROUGH TECHNOLOGY AND INNOVATION

The knitting industry, whether flat knit or circular knit, is driven by integrated capacities from knitting to garmenting. Bangladesh has taken leadership position in this segment for long and established a reputation of being upfront in staying at the position, investing in technology to continue as frontrunners in this segment. At ITMA, many Bangladeshi companies were seen at the various booths seriously exploring the new technologies and also understanding how the same would be able to add value to their business. It is notable that most of the technology providers too were excited to interact with the teams and as one technology provider said, "Today, the commitment to invest in automated and superior technology is coming from the South-east Asian countries, and Bangladesh in particular."

KNITWEAR SECTOR GOING FOR MAJOR AUTOMATION

The role of knitwear sector in Bangladesh's growth and development as a global leader in the apparel sector is immense! Registering momentous growth in last couple of decades – the category not only contributed significantly towards employment generation, women empowerment and poverty alleviation – it also played an important role in enhancing the country's GDP. In the process, the segment has pushed woven into second position in domestic value addition, which for the knitwear sector stands at an overwhelming 75 per cent, contributing 10 per cent to the country's GDP in terms of value addition.

Today, the knitwear sector, which can broadly be divided into two categories – circular knit (having fabric gauge of 12 to 22 and used for manufacturing T-shirts, lingerie, etc.); and flat knit (gauge ranging between 2 to 10 and used for manufacturing sweaters) – have in tandem been making steady progress in the country. Keeping pace with the growth, entrepreneurs have also strengthened backward linkages. Besides increasing their stitching capacities significantly, knitwear manufacturers have invested in allied industries to augment the overall capacity of the sector so much so that Bangladesh now scores over its competitors both in terms of pricing as well as product quality; thanks to industrial agglomeration, vertical integration and technological innovations.

While in circular knits the development has been more in terms of increased capacities, the flat knit has witnessed major developments in terms of technology. Aiding the growth of flat knit is the shift of not only the jacquard business from China to Bangladesh but the sweater business as a whole, due to rising labour costs and scarcity of skilled operators for linking operations in China. The trend has also encouraged many new companies to enter sweater manufacturing in the hope of replicating the growth revolutions of the late 90s. The advent of automatic flat knitting machines have further added an impetus with host of manufacturers engaged in the process of installing their 2nd or 3rd batch of automatic flat knitting machines.

"The kind of capabilities the country has for sweater manufacturing is difficult to be replicated by any other country, as not only is the investment huge but also the expertise and support system required have not kept pace with time in other countries," opines Saiful Hoque, Chairman of Sky Apparels, a comparatively new entrant in the sweater industry, who has installed first set of 100 flat knitting machines from STOLL in December 2013. Another set of 100 automatic flat knitting machines are ready to join the automatics from STOLL shortly.

Another sweater giant, Sonia & Sweaters has already installed 100 Shima Seiki machines and is committed to having 1,000 such machines in a phased manner over the next five years. "Profits are shrinking, labour is becoming expensive and buyers are looking for innovation. The only way to tackle these challenges collectively is to invest in automation," reasons Enayetuddin Md. Kaiser Khan, MD, Sonia & Sweaters Limited. He also adds that sweaters are becoming an important category in Bangladesh as now buyers get not only basic hand flats, but intarsia and other high-end knit options too. Having created a niche in 12 and 14 gauge sweaters, Sonia & Sweaters caters to brands like Marks & Spencer, Kenneth Cole, Gap, Metro Group, Carrefour Group, George and Betty Barclay Group.

Following the automation trend, US $ 60 million Crown Fashion & Sweater has installed 126 automatic flat knitting machines from STOLL and is looking to follow it up with more. "Since the programming system of the machines from both the brands (STOLL and Shima Seiki) are completely different, I wouldn't like to change the brand as we have a huge team of sampling operators and knitting machine programmers who are already trained on STOLL machines," shares Rezaul Ahsan, Chairman of Crown Fashion & Sweater, who also has two yarn dyeing facilities with a capacity of nearly 25 tonnes per day which cater to 75 per cent of the sweater manufacturers in the country. Manufacturing sweaters in gauges ranging from 3 to 12, the company majorly caters to the European market and is working with brands like H. Obermayer Gmbh, Karstadt Warenhaus AG, Distra Warenhandels, Gruppo Coin (OVS), Primark, Okaidi and Vegotex International.

A France-Bangladesh joint venture company L'Usine Fashion added 186 automatic flat knitting machines from Honkima and Juinn Long, replacing its 560 hand flat knitting machines in 2014, which along with Sky Apparels is now set to invest for heavy knit sweaters following the heavy gauge trend. Incidentally, STOLL recently introduced a heavy gauge knitting machine at ITMA which gives the look of a hand-knitted sweater in anticipation of the growing trend. "Basic products are necessary for sustenance, as they offer lower margins but bigger quantities; while the small order design-oriented products are for making profits. I don't know about others but this is my way of operating and with the new machines coming in from STOLL (for heavy gauges) and Cixing (for finer gauges) in our two upcoming factories, my share of jacquard and intarsia sweaters will increase and with it the FOB of my products," shares Prabir Saha, Managing Director of L'Usine Fashion.

Many companies are looking at major growth having invested in automation in the last few years. Investment in 100 automatic knitting machines from STOLL is now a two years old investment for the Crony Group and gives confidence that the turnover will touch the mark of US $ 120 million in 2016. Mauritius-based Floreal Knitwear Group, which produces 5.5 million sweaters per year, plans a big expansion in Bangladesh where it already produces 1.5 million pieces. "Our strategy is to double our production volume in Bangladesh within two years and produce 30,00,000 pieces there," informed Jean-Baptiste de Spéville, Chief Executive Officer of Floreal. The group has injected Mauritian Rupee 50 million (US $ 1.65 million) in the expansion project. Floreal specializes in manufacturing sweaters of cashmere, lambswool and merino.

Driving the trend of automation in Bangladesh, which started around year 2010, are companies like Shima Seiki and STOLL as manufacturers in overwhelming numbers and which opted to build capacities to produce high-end jacquard and intarsia sweaters, a speciality of these two machines. The reason for such major investments in the jacquard market was the expected flow of business to Bangladesh from China. However, despite the popularity of Shima Seiki and STOLL, companies from China like Cixing, Juinn Long, Gosheng and Honkima have not failed to capture substantial market shares; thanks to their single system machines, which apart from being cost-effective also serve as a replacement for the hand flat knitting machines and the first contact point to automation before graduating to high-end machines.

Functionality, Fashion and New Applications the Directions

Among the key segments that saw major developments and even a few breakthrough technologies at the recently concluded ITMA was 'knitting'. The wide range of knitting solutions – from flat knits to warp knitting to circular knits to seamless technology – all were on display for the many visitors who thronged the corridors over the week-long event. Asian companies, mostly from the Indian sub-continent, were seen in large numbers and it was not surprising that for many technology manufacturers, this region is a focus market, with both strategy and machines being made to penetrate the potential this region presents today.

Each segment in knitting had some new machines to show, but the biggest 'disruptive' innovation was in circular knitting with two companies putting forward their version of a knitting machine wherein the spinning aspect was integrated. While Pailung from Taiwan previewed to some select customers a circular knitting machine that directly uses sliver instead of yarn for knitting fabric, Mayer & Cie presented its 'spinitsystems' that is a 3-in-1 concept where spinning, cleaning and knitting happens on one machine. On the new machine, knitwear is manufactured not from yarn, but straight from the fibre roving, significantly shortening the production process and also reducing the capital expenditure required for a knitting unit.

Though Pailung was still keeping the technology under wraps, Mayer & Cie confidently talked about the advantages and also applications of the technology. The brand also claims that the versatility of the machine is such that besides natural fibres like cotton, the machine can also knit synthetic fibres from the short staple area, such as viscose, modal or polyester. Not forgetting the sustainable importance of technology, the 'spinitsystems' claims 30 per cent lower CO2 emissions running at 25 RPM to produce 20 kg/hour of knitted fabric.

Another interesting development in circular knits came from the Korean company Keumyong, which after two years of dedicated R&D in association with Nike, has introduced a double knit electronic full jacquard with rib-mesh machine (KCMJD-169S). This new technology allows circular knits to enter new product markets like shoe fabrics. The electronic individual needle selection and unlimited patterning variety realize the quality of jacquard and versatility of fabrics. "Though both flat knitting and warp knitting are presenting options for the shoe industry, circular knitting is much faster and changing colour here is so much simpler. We already have an order for 1 million pair of shoes from Nike on this machine and any company which invests in the technology does not need approvals for the same," claims C.W. Son, Office Manager (Seoul), Keumyong Machinery.

Displaying its version of the double knit double selection electronic jacquard knitting machine, under the Knitel series, Pailung was promoting the machine for home textile applications, more so for knitted mattress, which is a growing trend today. "The Pailung brand has been doing exceedingly well in India and Bangladesh as it is a good balance of high-end reliable technology and price. R&D is the core growth engine of the company and the industry players are offered new options to expand their business on a regular basis," said M. M. Elango, CMD, Yuvraj Merch-X, agents for Pailung in India. Also joining the Pailung range of knitting machines are a high speed range of circular options from Vanguard, USA. Focused at increased productivity, VPL (Vanguard Pai Lung) is not looking at only increasing speed, but also tackling all issues that arise out of increasing speed. "Our machine knits about 2,000 pounds a day at 60 RPM, with controlled vibration and heat. We don't have many machines and we knit only jerseys and fleece, but we dominate our region on speed. Machines can be customized from a select range of frames, diameters and gauges," informed Mike Moody, VP, Product Development, VPL.

Warp knitting is another area that is developing fast and companies like Karl Mayer have taken development to the stage where one will not have to wait for too long to see warp knitted fabric replacing weaving in manmade fabrics. Produced much faster, sturdier and endowed with stretch, the fabric produced are now being extensively used in lingerie and sportswear, and as automotive textiles and shoes fabrics. The HKS series on display at the fair attracted many visitors. Representing Karl Mayer in India, A.T.E. Enterprises was very happy with the response. "Weaving has always been a very capital-intensive industry and now high-end elastic fabric with performance value can be made anywhere, putting an end to limitation in fabric suitable to performance-driven segments like sportswear and shoes," said Vikram Chaurasia, Deputy GM, Textile Engineering – Fabric Forming, A.T.E.

While warp knitting was focusing on performance, the two big companies in flat knitting were concentrating on fashion as the USP. Shima Seiki had put on display a complete new collection of fabrics/products created from their machines, along with three upgraded versions of the WHOLEGARMENT concept which is celebrating 20 years of innovation. The designs were highly complicated and the final products were being lapped up by many European customers. “The technology we are displaying here is more suitable to the European customers who are looking for new inspirations in designs,” said Ikuto Umeda, Executive Director, Shima Seiki Mfg. Ltd. For the company, while Europe is the high-end market, the big mass market is in Asia, particularly India and Bangladesh.

There is no doubt that Bangladesh is the second largest market for flat knitting machines today after China, and the companies are going all out to capture the growth. “In last three years we have sold 5,000 machines to Bangladesh, and though the market is a bit shaky now because of huge investment in automation without prior preparation of trained manpower, the market should continue to grow, albeit a bit slower, over the next two years before picking up speed again,” predicts Thomas Hoffmann, Area Sales Manager, STOLL who is looking after the Bangladesh region. Hoffmann further shares that there are three major factors that are responsible for the growing demand for automation in Bangladesh, the first being the increasing wages in the country, the second being the need to give more consistent quality, and finally for the simple reason that it is easier to be a compliant factory with less workers!

Contrary to common perception, automated machines from branded companies are not direct replacements for hand-flat machines, being commonly used in Bangladesh. “Hand-flats are being replaced by Chinese machines which are semi-automated and give better productivity, but still translate into one machine per person. Our machine replaces around 7 hand-flats and one person is operating 4 machines on an average; this means that one machine replaces about 23 people,” argues Hoffmann. He adds that in China the ratio is 8 machines per person and in Turkey it is as high as 12 machines per operator. Of course productivity per person is not only about skill levels of the operator but also the quality and type of yarn being used. The company had on display a new coarse knitting machine (STOLL CMS series) with the ability to give authentic hand knitted look.

Keeping pace with fashion trends, STOLL has expanded its gauge range further, adding gauge E1.5 and gauge E2.2 for the extra-bulky knits and extra-chunky yarns that previously demanded hand knitting. Gauge scalability now extends stitch optics from gauge E1.5 to gauge E5; yarns down to 0.7 nm can be used. Moreover, the industrial knitting machine offers countless design options: jacquard, intarsia and structured knits – combined with shaping at tributes in Fully Fashion technique. The design applications extended beyond apparel and included accessories and interior design.

Earlier this year STOLL announced a limited partnership with Shanghai-based Shang Gong Group (SGG) which is focusing on sewing and textile welding. Management at STOLL emphasizes that all current corporate values – i.e. innovation, quality, service, support and responsiveness – will remain key priorities. This partnership supports future growth and sustainability and is the basis for further expansion of products and solutions. STOLL will benefit from additional know-how and an even broader presence in Asia. "SGG is a strong international partner which deeply understands the textile industry.

The exhibition was a showcase for the company's latest selection of computerized flat knitting machines, which gathered great interest from visitors. The new flagship MACH2XS WHOLEGARMENT knitting machine with SlideNeedles mounted on four needle beds and spring-type sinkers, the compact SWG-N2 series WHOLEGARMENT knitting machine with capability for industrial textiles, as well as the SRY183LP with loop pressers that produce unique knitwear with woven textures, is now available in ultrafine 16 gauge. MACH2X achieves a maximum knitting speed of 1.6 metres per second. The R2CARRIAGE system furthermore permits quicker carriage returns for higher efficiency per knitted course. Split Stitch technique also allows efficient knitting by eliminating empty courses. Additionally, MACH2X123/153 features 18L gauge capability whereby a special large-hook version of the SlideNeedle is used for knitting 15-gauge fabrics at 18-gauge needle pitch for very high productivity when knitting ultrafine gauge garments. The tighter needle pitch also allows for a tighter fabric especially for ribs, contributing to higher-quality WHOLEGARMENT items. Additional improvements include a control unit that is built-in to the machine for simplified shipping and installation, easier maintenance, and more efficient use of space. A USB memory interface and a backup power supply are standard. MACH2X also features a lower overall height for improved access to top tension units.

The Keumyong booth at ITMA was a beehive of activities as the Korean circular knitting machine manufacturer had developed a jacquard model, made specifically for the production of footwear – KCMJD-169S. The double knit electronic full jacquard with rib-mesh machine is a multipurpose double knit machine with a wide range of stitch options that comes in two diameters of 30 inches and 34 inches with gauge options from 12-18. The machine runs at an RPM of 12-15 and has a 3 position needle selection by WAC (Knit, Tuck and Welt). The company is very upbeat of the immense potential of the machine and feels that it will change the way sports shoes are being made today.

PAILUNG THRIVES ON WIDE RANGE OF KNITTING MACHINES

The Taiwanese manufacturer of knitting machines is one of the most popular in Asia, with good market share in both India and Bangladesh. The company has machines for all applications from basic fabrics to specialized fabrics for use in home furnishing, sportswear, automotive industry, healthcare and industrial purposes, under three categories of machines – Knitel, Weevknit and Rediknit. The knitting machines are custom built for brands, designers, textile manufacturers, and contractors in the fashion, home, healthcare, and environmental textile industries.

The company believes in the collaborative approach to R&D and works closely with customers to build better knitting machines for current industry needs. They also claim to be fast on the delivery of customised machines, allowing clients to keep pace with most challenging needs and aspirations. At ITMA, among the many machines on display, the mattress knitting machine was drawing good attention from visitors. PL-KDDSCJ is a thoroughbred model with the latest technology and refinement, which can present very clear patterns of multiple colour designs on the front face without colour moving to the backside.

Knitting Technology Developments

Courses, wales, loops, gauge, warp, weft, raschel, jersey, interlock ... just some of the terms familiar to people working in the knitting industry. Knitting technology comes in a variety of types and sizes and may be used to make all sorts of products from small vascular heart grafts (see "Prototype Graft Designed To Replace Damaged Heart Vessels Shows Promise In Cell Study", TW, January/February 2021), to shoe uppers and apparel fabrics all the way to large-scale bedding components and anything in between that requires comfort, stretch and seamless shaping, among other attributes.

Since ITMA 2019 — ITMA typically paving the way for new innovations — and despite difficulties in the machinery market because of the pandemic, knitting technology manufacturers are still innovating and helping their customers provide value. Here is a look at a few new products available to knitters.

Mayer & Cie

After completing field tests, Germany-based Mayer & Cie reports its OVJA 2.4 EM circular knitting machine now is ready for the market. The fully electronic model is designed for mattress cover fabrics with high output and a wide pattern variety. This ready-for-market version builds on the model introduced during ITMA 2019 with the addition of a new thread fluctuation control system positioned on every second feeder. A controlled air stream maintains constant yarn tension especially at high revolutions per minute, which helps avoid thread loops and dropped stitches.

"With a feeder density of 2.4 the OVJA 2.4 EM tallies precisely with the current trend toward higher feeder densities that make higher productivity possible," said Thomas Zizmann, area sales manager at Mayer & Cie. "The OVJA 2.4 EM produces up to 30 kilograms of fabric per hour. That makes it demonstrably the most productive machine of its kind in the world, as shown by its speed factor of 950."

Relatively new to the mattress market, knitted fabrics have taken off over the last 20 years according to Mayer & Cie in part because of their elasticity and flexibility. However, the market is increasingly competitive, and Mayer & Cie. aims to add value for its customers — mostly premium manufacturers in the United States and Belgium in this sector — by increasing the production output of its mattress machines.

Mayer & Cie also offers the OVJA 1.6 EE/2 WT double jersey jacquard machine for multi-colored designs and microstructure elements including tuck structures, spacer fabrics and double-knit fabric with a lay-in thread. The machine features three-way electronic needle selection in the cylinder and two-way technology in the rib dial. When special needles are employed, the machine can knit yarns up to 1,200 denier in gauges as coarse as E16. Applications include mattress covers, upholstery, transportation seat covers, shoe uppers and outerwear.

In keeping with the digitalization trend, Mayer & Cie. also offers a cloud- and web-based digital solution for the knitting industry via its knitlink technology. knitlink records data from each machine and provides operators with a visual record of the data on a personalized dashboard.

Karl Mayer

On the business side, in 2020 Germany-based Karl Mayer Textilmaschinenfabrik GmbH acquired H. Stoll Ag & Co. KG., an established name in the flat knitting arena. The purchase broadens Karl Mayer's product offerings to include both warp knitting and flat knitting solutions. Stoll now operates as an independent business unit within the Karl Mayer corporate group.

In the United States, the Stoll and Karl Mayer North American subsidiaries were consolidated into one location in Greensboro, N.C. An investment at the site, scheduled for completion in March 2021, includes a state-of-the-art textile development and visitor center, which offers demonstrations, training, collaboration opportunities with industry partners, prototyping and an assembly operation for both Karl Mayer and Stoll technologies. Stoll's core team in Greensboro offers business and technical management, textile design and product development, machine programming and customer service.

"With the new development and visitor center and the synergies that come with combining our capabilities, we offer textile producers and all industries that can imagine the use of textiles an extremely strong resource in North America" said Tony Hooimeijer, president of KARL MAYER North America.

The company also has created a Textile Makerspace platform intended to help connect the textile industry with new technologies to encourage innovation. One focus, Textile Circuit, looks at smart textiles and wearables, and projects are underway at the company's headquarters in Obertshausen, Germany. The Greensboro location also offers a Makerspace area with a modified standard machine and competent service technicians, according to the company, where U.S.-based customers can develop and implement ideas for e-textiles with support from Sophia Krinner, textile technology product developer, who was charged with setting up Textile Circuit. "Customers can come to us with their ideas and work with us to try out different ways to make their ideas a reality," Krinner noted.

On the technology side, Karl Mayer offers some new technologies for the knitting sector. Its RD 7/2-6 EN double needle bar raschel machine offers up to 30-percent higher speed compared to its predecessor the RD 7/2-12 EN, according to the company. First introduced in 2019, the machine can knit 3D fabrics using a trick plate distance of 2 to 6 millimeters. Its EN pattern drive and maximum shogging movement of 1 inch allow patterning variety with up to 12,000 stitch repeats. The machine is available in working widths of 138 and 77 inches in gauges E 18, E 22 and E 24. Fabrics may be used for footwear, clothing and mattress applications.

Karl Mayer's HKS 3-M ON was one of the first machines connected with the company's digital solutions. The machine can directly download pattern data from the KM.ON secure cloud and does not need to use pattern disks, which allows for easy pattern changes found with EL gears at the speeds associated with N pattern drives. The machine recently was issued in a wider working width of 300 inches for wider or side-by-side production, and is up to 15 percent faster than its predecessor, according to the company.

Karl Mayer offers three interchangeable options for purchasing lappings for the HKS 3-M ON — SwapKnit 36 and SwapKnit 36 Flat for designs up to 36 stitch courses; as well as a variant developed for pattern repeats of more than 36 stitch courses. Patterns for the machines are downloaded from Karl Mayer's Webshop Spare Parts and loaded on to the knitting machine for fast and flexible pattern changes. According to Karl Mayer, the hybrid machine combines the patterning possibilities of a typical N and EL pattern drive in a single machine.

Shima Seiki

In the early days of the SARS-CoV-2 pandemic, Japan-based Shima Seiki Mfg. Ltd. released free knitting data for masks that could be produced using a variety of its knitting machines including the Wholegarment® technology. The goal was to help Shima Seiki machine owners who were looking to switch production to masks to alleviate the shortage of surgical masks as quickly as possible. The patterns were for 3D form-fitted masks featuring integral ear loops that could be knit in one step with no further sewing required. The designs are available at Shima Seiki's user website located at shimaseiki.com/user/samplesearch.

Officially launched in September 2020, Japan-based Shima Seiki Mfg. Ltd.'s yarnbank™ online web service allows users to search and view the latest yarns from a wide variety of companies around the globe, and then download the yarn data for use by fabric simulation and virtual sampling on the Shima Seiki SDA®-onE APEX4 design system. The goal is to give designers the ability to create accurate representations of garments that may be created using yarns that are available for production.

Existing Shima Seiki APEX users can register to use yarnbank free of charge. In October 2020, the company also launched a subscription-based design software option named APEXFiz™. This offering breaks with Shima Seiki's tradition of offering only all-in-one proprietary hardware and software packages and allows customers to install the software — available in five different configurations: Design Jr.; Design-Standard; Design-Knit; Design-Weave; and Design Pro — on their individual computers. APEXFiz users also have access to yarnbank.

Shima Seiki reports it "has traditionally promoted design simulation and virtual sampling as an essential part of its 'Total Fashion System' concept wherein virtual samples replace physical samples in an effort to reduce time, cost and materials wasted in the sample making phase, further realizing overall efficiency and reduced waste for a sustainable manufacturing supply chain."

Yarn companies can register their products on yarnbank under a number of different plans. The advantage to yarn manufacturers, according to Shima Seiki, is the ability to "present their yarns directly to their customers." The yarnbank technology can save time, costs and materials for yarn manufacturers by reducing the dependence on traditional sample books for product promotion.

Shima also recently released the SHIMANAVI™ e-learning system that provides online training for its APEX series users. The flexible system offers support at the user's pace when and where it is convenient. Several of the courses are available in multiple languages.

Comez

Some recent technology introductions from Jakob Müller Group company Comez Internationl S.r.l., Italy, include the DNB/600-2B, DNB EL-32 and 800 ACO/EL. The DNB/600-2B is a high-speed mechanical double needle bed warp knitting machine designed for mass production of items including ear loops for masks and sportswear accessories. The low stitch density required for such articles and high operating speed of the machine equate to very high productivity, according to Comez. Equipped with two mechanically controlled guide bars, the DNB/600-2B can process all types of yarn into double face fabric structures. A compact design and reduced working width result in fast changeovers.

The company's DNB/EL-32-12B, was first introduced at ITMA 2019. The electronically controlled warp knitting machine is designed for sample production using any yarn in all possible gauges on a single, compact machine. Items that may be produced using the DNB/EL-32-12B include technical textiles, medical textiles, shoe uppers and apparel textiles. The machine's 12 guide bars and electronic long-throw device with a maximum stroke of 560 millimeters allow more advanced technical fabrics to be produced such as reinforced fabrics, fabrics with continuous weft insertion and technical furnishing fabrics.

Comez's 800 ACO/EL crochet knitting machine features compound needles, electronically controlled yarn tension and stitch density regulation. Comez reports the machine is suitable for the production of elastic and rigid fabric and tapes, and is ideal for products requiring a lock-stitch structure such as medical and orthopaedic textiles, closed-loop Astrakhan-style fabrics, furnishing ribbons and tapes, and sportwear ribbons, among other articles. The machine is equipped with eight weft bars and is available with an 800-mm working width.

RECENT DEVELOPMENT OF CIRCULAR KNITTING MACHINE

Circular Knitting Machine

Circular knitting machine is widely used throughout the knitting industry to produce fabric. This machine can be built in almost any reasonable diameter and the small diameter of up to five, which are used for wear.

Seamless knitting technology

Seamless garment knitting can be achieved either on the circular knitting machine or flat (V-bed) knitting machine. However, seamless circular knitting machines differ from seamless flat knitting machines in that seamless circular machines create only a single tubular type of garment such as those produced on Santoni machines.

Seamless knitting machines can create more than one tube and join the tubes together on a machine. The complete garments knitted on circular machines may also only need a minimal cutting operation.

In addition, seamless circular machines require different diameters to make major changes in garment size, whereas seamless flat machines can adjust to different garment sizes on the same machine. Consequently, seamless knitting on circular machines is not true seamless knitting. It should be mentioned that knitting on V-bed seamless machines produces truly seamless garment since they do not require any cutting or sewing. In recent years, Santoni has developed a four-feed single-jersey electronic circular machine, which enables the creation of a shaped garment by reciprocal movement.

To get a higher quality knitted garment, it is crucial to control the manufacturing functions. It is critical for designers and manufacturers to communicate effectively in order to create successful new products in the knitting industry. Designers complain that the designs that they specified are not accurately created, while technicians are of the opinion that the designers do not understand the technical problems in knitting feasibility. It has been proposed that one way to overcome the communication problem between designers and technicians would be the use of intelligent CAD systems. The CAD system gives designers and manufacturers the opportunity to specify and evaluate their design more precisely without requiring great time investment and technical expertise. Diverse computer-controlled systems including CAD/CAM controlled machines have been developed to facilitate communication.

Companies have been offered new types of CAD system, which use two different monitors including a technical window and a design window for designers and manufacturers, who require different information for the same design.

Ultra-fine gauge knitting machines

Fine knitted fabrics known as 'second skin' are produced on fine gauge circular knitting machines. Cotton, polyester and viscose yarns of 90 to 120 Ne are applied to produce fine circular knitted fabrics. The appearance of these fabrics is similar to woven fabrics but they are more flexible.

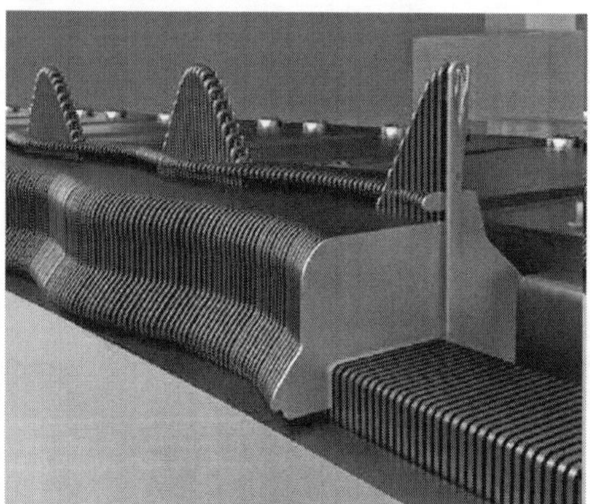

High-grade circular knitting cylinders provide the ideal complement to the company's premium range of needles and system parts to create the perfect knitting system from a single reliable supplier. It is only with the guarantee of a consistently high standard of component quality and outstanding durability that circular knitting machines are able to reach their full potential for high-performance operation on the factory floor. Ensuring the pinpoint precision of individual elements in existing machines not only simplifies the workflow but also improves capacity utilisation in production.

Fine gauge automotive fabrics, known as woven-like fabrics, are produced on fine gauge double cylinders in Pai Lung ultra-fine gauge machines. Pile technology is applied to these machines, where the fine gauge of loop makers is adjusted near to the sinkers.

LOOP TRANSFER TECHNOLOGY IN CIRCULAR KNITTING MACHINES

Stitch transfer is an important operation, as a higher number of transfer modes means more possibilities of making structured designs and shaped fabrics in which the stitch is transferred from one cylinder to the other or within the same cylinder.

The production of high-quality knitted fabrics – i.e. of homogeneous appearance as a result of a smooth knitting process and the absence of holes and barring – essentially depends on the application of certain technical solutions.

As for the type of needle used, Italian machines incorporate latch needles, which operate according to the drowned butt principle. This kind of needle remains in an idle position with its heel completely drowned in the needle bed groove without being involved in the action of cams, and retains the loop, which in this case is not subject to strain. Figures 7.7 and 7.8 demonstrate novel loop transfer elements of these circular knitting machines.

Pile and sliver insertion mechanism in circular knitting

A special sliver knitting process locks individual fibres directly into a lightweight knit backing, allowing each fibre to stand upright, free from the backing, to form the soft pile on the face of the fabric. This makes comfort pile fabrics softer, warmer, more drapeable and more resilient than fabrics made from yarns.

Each fabric originates from premium loose fibres. These fibres include high-tech microfibre acrylics, polyesters and mod-acrylics specially developed for fabric, along with natural fibres such as wool. Each blend is chosen for its specific end result. By engineering the fibre mix, an incredibly wide range of colours, density, weight, patterning, texture and performance features can be produced in comfort knit pile fabrics.

After knitting, the pile fabric is sheared to the desired height. It is then put through a series of technical finishing processes specially developed by this kind of knitting process to control the surface texture and special characteristics of the final fabric.

Advancements in circular Knitting Machines

Following are the advancements in circular Knitting Machines:

Introduction to Seamless Knitting:

We can achieve the seamless garment knitting either by using the circular knitting machines or by using the v-bed flat knitting machines. Seamless circular knitting machines are different from the seamless flat knitting machines. By using the seamless circular machines, we can create only a single tubular type of garment whereas when we use the seamless flat knitting machines, we produce more than one tube which we have to join together on a machine. The complete or single tubular type of garment which we knit on the circular knitting machines requires a minimal cutting operation.

Moreover, if we want to make any significant change in the size of the garment, then we require different diameters in case of the seamless circular knitting machines whereas in seamless flat knitting machines we can change to the different sizes of the garments on the same machine. Therefore, the seamless knitting which we achieve on the circular machine is not the true seamless knitting. On the other hand, V-bed seamless machines produce the truly seamless garments because they do not need any cutting or sewing of the fabric.

Automatic Seamless Circular Knitting Machine:

It is essential to control the manufacturing functions of the machine to manufacture an excellent quality knitted garment. For successfully producing new products in the knitting industry, it is necessary that the designers, as well as the manufacturers, effectively communicate with each other. Most of the times, the designers complain that the technicians do not accurately create the designs that they specify, whereas the technicians are of the opinion that the designers do not understand the technical problems that they have to face in the knitting feasibility. However, to solve the communication problem between the designers and the technicians, it's always better to make use of intelligent CAD systems. When we use the CAD system, then the designers, as well as the manufacturers, can specify and evaluate their designs more correctly without requiring technical expertise and taking much time investment.

To facilitate communication between the designers and technicians we have diverse computer controlled systems such as CAD/CAM controlled knitting machines. At present, knitting machine manufacturing companies also have a facility for several new types of CAD systems. The new types of the CAD systems use two different monitors as well as have a design window for designers and a separate technical window for the manufacturers because both of them require different information even for making the same design.

Ultra-Fine Gauge Knitting Machines:

We use fine gauge circular knitting machines for producing the fine knitted fabrics which we call as the 'second skin'. For creating the fine circular knitted fabrics we commonly use polyester, cotton and viscose yarns of 90 to 120 Ne. The fine knitted fabrics have an appearance similar to that of the woven fabrics, but they are more flexible as compared to the woven fabrics.

Ultra-Fine Gauge Elements Of A Circular Knitting Machine:

To complement the company's premium range of needles and system parts, we have high-grade circular knitting cylinders. High-grade knitting cylinders help us in creating a perfect knitting system from a single reliable supplier. Circular knitting machines can attain their full potential for giving high-performance operation on the factory floor only because of the consistently outstanding durability and high standard of component quality. We always ensure to install every component of the existing knitting machines with utmost precision. The precise installation of the individual element not only improves the capacity utilisation in production but also simplifies the workflow.

Pai Lung ultra fine gauge knitting machines consist of the fine gauge double cylinders which we use to produce the fine gauge automotive fabrics called as woven-like fabrics. Ultra-fine gauge knitting machines make use of the pile technology, and the fine gauge of loop makers is present near to the sinkers.

Loop Transfer Technology In Circular Knitting Machines:

Stitch transfer is an essential operation in the process of knitting. Higher the number of transfer modes, more are the possibilities of making structured designs and shaped fabrics. For creating the structured designs, we transfer the stitch within the same cylinder or from one cylinder to the other.

The high-quality knitted fabrics have a homogeneous appearance, and they don't have any holes and barring in them. To produce high-quality fabric, we require a smooth knitting process as well as the application of certain technical solutions.

Concerning the needle used, the Italian knitting machines have latch needles incorporated in them, which operate by the drowned butt principle. In Italian machines, the needle remains in an idle position. The heel of the needle remains wholly drowned in the needle bed groove and do not involve in the action of the cams. The heel also retains the loop, which in this case is not subject to any strain.

Pile And Sliver Insertion Mechanism:

The sliver knitting process locks the individual fibres directly into a lightweight knit backing. The sliver insertion mechanism makes the fibres free from the backing and allows them to stand upright; thus forming a soft pile on the face of the fabric. This technique helps in making the comfort pile fabrics softer, more resilient, warmer and more drapeable as compared to the fabrics made from the yarns. The fibres which we commonly use include high-tech micro-fibre-acrylics, mod-acrylics and polyesters along with that of the natural fibres like wool.

Loop Transfer Tool:

We carefully choose each blend to gain the specific results. By engineering the fibre mix, we can quickly produce the comfort knit pile fabrics in an incredibly wide range of colours, density, weight, texture, patterning and performance features.

Needle with Spring for High-Speed Loop Transferring:

After the process of knitting, we shear the pile fabric to the desired level of height. The fabric then undergoes a series of technical finishing processes specially developed to control the special characteristics as well as the surface texture of the final fabric.

7

CABLE KNITTING PATTERNS

Cable knitting is a style of knitting in which textures of crossing layers are achieved by permuting stitches. For example, given four stitches appearing on the needle in the order ABCD, one might cross the first two (in front of or behind) the next two, so that in subsequent rows those stitches appear in the new order CDAB.

Cables like the ones in the swatch shown are made with knit stitches, and are surrounded on each side by purl stitches (reverse stockinette stitch) to make them pop off the fabric. Cable patterns can also involve different combinations of knits and purls (a little more about this later when we talk about more complex cables), but for these tutorials we'll be working with knit cable stitches.

For this swatch, we worked all the cables in one color, and then we worked a bit with two colors to better illustrate what's happening. You normally will not be working cables in two colors (you can, but it's more advanced) so the colors are only for you to see the front/back or left/right stitches.

Use the cable needle that is right for you. I found that I kept dropping the little short/straight needles made specifically for cabling, so I tried a regular sock-sized double-pointed needle. Worked like a charm, because my fingers already knew how to wrangle that one. But that's just me. There's many different choices out there when it comes to cable needles. Try a few until you find the one that works for you.

Corollary: Try cabling without a cable needle. It sounds impossible, but you can learn to manipulate the cable crossings without the extra needle. There are many cable knitters who swear by this technique! I did the first end of the scarf with a cable needle, and then I tackled the second end without one. The effort it took to learn which stitches went where was well worth it, because now I feel as though

I am understanding cables instead of just knitting them by rote. We have a list of online tutorials in our Techniques section.

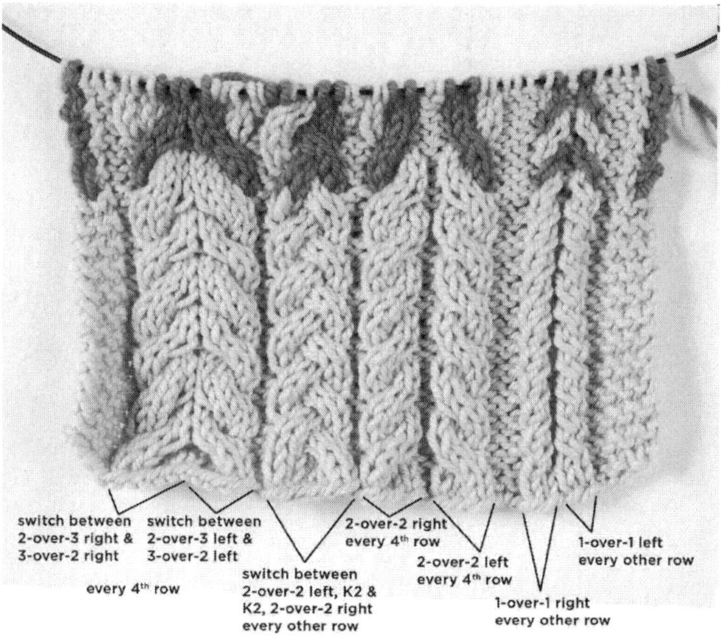

Don't pull on the "held" stitches too hard. When you are holding the cable stitches off to the front or back of your knitting, don't pull them too far away from the rest of your knitting! Too much pulling will distort the stitches in the area of the crossing. Keep them as close to the main knitting as possible in order to help keep the tension and texture of your cable stitches even.

WHAT THE BACK LOOKS LIKE

Watch what you are doing. Until you are comfortable with cables, cable knitting is not the time to multitask! I made several whopping mistakes because I was trying to knit the cables in a restaurant whilst talking with friends. Bad idea. (Lots of ripping out ensued.)

Steam-block Cables wrong side up. This may seem obvious, but just in case: If you are using a steam-iron to block your cable knitting, do it with the WRONG side of the cables facing upwards, or you will flatten all your nice intentionally-bumpy cables. Don't press down--keep the iron just a little bit above the fabric! And try using a pressing cloth to further help minimize the flattening factor.

You asked for it...Lynn G. asked if I would be willing to post a photo of what the reverse side looks like. There you go, Lynn!

Cables are a wonderful way to add interest and texture to your knitting. The very basic cable looks a lot like a rope in the middle of a field of purl stitches. However, there are hundreds of patterns out there that will teach you how to make large, beautiful, complex cables. Looking at a cable on your sweater makes it seem confusing, but it's not at all.

All a cable is is a group of stitches that are knit out of order on your needle. To get your stitches out of order, you put them on what is called a cable needle and hold them to the front or back of your work. After knitting the next few stitches, you slide the stitches on the cable needle back onto your left needle and knit them.

Step 1Gather Materials

For this instructable, I'll be knitting a simple swatch with two identical cables. I got the "climbing cable" design from this book:

Here's what you'll need:

knitting needles

yarn

cable needle

scissors

optional:

cable pattern

stitch markers

Amy's free knitting pattern allows you to make a custom-sized cabled headband that ranges from kids' sizes small/medium to adults' sizes medium/large. The measurements include circumferences between 11 inches to 20 inches and widths between 3 inches to 3.25 inches. This gorgeous, wide, and snuggly-fit cabled headband is knitted with a 50% Angora and 50% Wool yarn called Classic Elite Lush using a size 5 (3.75 mm/12-inch circular needle), a cable needle, and a tapestry needle.

Published on Knit Picks website, this free knitting pattern makes a brilliant, wide cabled headband with a lattice cable design that resembles a woven design. The recommended yarn is one 50-gram skein of Elegance York (70% Baby Alpaca and 30% silk) and size US 6 (4 mm) straight or circular needles.

Katie has designed a simple pattern for an exquisite "Anthro Inspired Plaited Headband" that resembles a thick braid that wraps around your head. The required materials include one skein of bulky weight yarn, size 10 knitting needles, and a cable needle.

Natalex has designed a knitting pattern for a wide, multicolored, cable knitted headband that fastens with a button. The materials needed include Lamb's Pride bulky yarn (two skeins of a different color), a size 10 circular needle, and a cable needle.

Available for purchase in Interweave Knits, Winter 2006 magazine at the online Interweave Store, is the "Cable and Bobble Headband" pattern designed by Carrie Bostic Hoge. This awesome bulky and wide, multicolored cabled headband with I-cord ties is knitted with one skein of Caron Felt-It yarn (100% wool), two size 10 ½ (6.5 mm) double-pointed needles, a cable needle, a tapestry needle, and an optional pom-pom maker. The finished size is 3 ¾ inches wide by 18 inches long.

Overall, the instructions for these cable knitted headband patterns are simple to master. The main concern is knitting these headbands with a snug fit that tailors to an individual's head size. The designer of the cabled headband "Nakiska", Amy Swenson, offers the advice of "measuring the circumference of the head and subtracting 3- 5 inches" since stretchiness increases with wear. Of most value, women, men, and teens will be pleased with the warmth that cable knit headbands provide during the winter.

Complex Cables

Most cables you encounter will be worked just the same as the ones we've done above. Some complex cables use two cable needles with more than two sets of stitches — we won't get into those today — but you should be able to follow even these tricky cables by working each step of the pattern as written. Most patterns use cables that have a set of stitches (could be 2, or 3 or 4…) twisting either right or left over another set of stitches, same as what we've already done.

The most common variation you'll find is a very simple one: purling the stitches in the back instead of knitting them. For this, you'll work the cable exactly as the instructions show above (either with or without a cable needle) but purl the stitches held to the back of your cables, which are the first stitches on a left twist, or the last set of stitches on a right twist.

Another complex type of cable you may find in some patterns is one that incorporates an increase or decrease into the cable stitches. If you know how to work the increase or decrease, and you know how to make the cable, then you'll be able to follow the steps and make the pattern!

In the middle of this swatch image is an example of another simple use of 2-over-2 cables. The braid style cable is made by twisting 6 stitches total in 2-over-2 cables, one cable twist per right-side row. To work this cable: on a right-side row, work the first set of 4 stitches as a 2-over-2 Left cable, then knit the last 2 stitches normally; on the following right-side row, knit the first 2 stitches normally, then work the next 4 stitches as a 2-over-2 Right cable. The way the stitches are cabled around each other is just like a 3-strand braid (think hair braid!): twist the outer 2 stitches up over the center 2 stitches, first on one side, then the other side, and just keep repeating.

Here is an example of asymmetrical cables — there are 5 stitches per cable twist, so first a 2-over-3 cable was made, then a 3-over-2 cable next time, so sets of 2 stitches are twisting around sets of 3 stitches.

And it's also an example of two cables worked side-by-side, with no purl stitches or plain stitches between then, which makes a broad, flat cable. Use the same techniques as above, but immediately one after another.

Here, the 3-over-2 Left cable has just been completed. This is worked exactly like a 2-over-2 Left cable, but with 3 stitches on the cable needle held to the Front instead of 2.

Then immediately after that Left cable, a 3-over-2 Right cable was worked. This is the same as the 2-over-2 Right cable, except after holding the 2 stitches on the cable needle to the Back, 3 stitches are knit for the front stitches of the right twist.

To continue on with the above cable pattern, the next right-side row will be knit across normally, and then the following right-side row will be cabled: first a 2-over-3 Left cable, then a 2-over-3 Right cable. You can see in the swatch how the sets of 2 stitches are yellow, and the sets of 3 stitches are blue, so those sets always remain grouped together as the cables twist around.

Now that you understand how cables work, let's take another look at that complete swatch! Can you see what's happening in each cabled column? See how as the cable twists travel up, the stitches twist around each other, either to the right or to the left, and you now know that those stitches are twisted in front by being pulled over the front of the other stitches.

Experimenting?

A couple of extra things for if you're NOT working from a pattern. These tips are unimportant if you're following a pattern, as the pattern will tell you exactly how to work each row. However, if you want to experiment with your own cable ideas, read on if you for a better understanding of the basic "rules" of cable construction.

First, the general rule of thumb for how often to work a cable twist is: work as many rows are there are stitches in the cable for each repeat. Round down for an odd number of stitches (unless you're working in the round, in which case there's no reason you need to work cables only on odd or even number rounds).

Here are some examples

For a 1-over-1 cable, there are 2 cable stitches involved, so work the cable twist on every 2nd row, which means every right-side row.

For a 2-over-2 cable, there are 4 cable stitches involved, so work the cable twist on every 4th row. This means after working a cable row, you'll work 1 wrong-side row, then 1 right-side row, then 1 more wrong-side row; repeat the cable row on the next right-side row.

For a 3-over-3 cable, there are 6 cable stitches involved, so work the cable twist on every 6th row. So you'll work 2 plain right-side rows between cable rows, making six rows total in the pattern repeat (3 right-side, 3 wrong-side).

For an asymmetrical 2-over-3 / 3-over-2 cable, round down to 4 stitches and work the cable every 4th row. Or you may want to try rounding up and working the cable every 6th row, and see the difference between the two options; maybe you'll prefer one over the other.

For the braided cable we did above, a cable is worked on every right-side row because they switch sides. So the first kind of cable — 2-over-2 Left over the first 4 stitches — is only being worked every 4th row, and the other cable, worked over the last 4 stitches, is being worked on the opposite right-side rows, also every 4th row.

Another tip to keep in mind is that cables pull in the knitting significantly, as the stitches pull across left and right to twist, so adding cables to a project will significantly affect your stitch gauge. For example, if you know that a hat in the same yarn weight that fits you well has 100 stitches around in stockinette, but you want to add some cables, you'll need to cast on more than 100 stitches to compensate for the change in gauge and elasticity. Try adding half as many stitches as there are in the cables; so, if you're adding five 2-over-2 cables (which use 4 stitches each), that's 20 stitches total in all the cables, so add 10 extra stitches. This math may or may not be perfect, depending on your yarn, density, specific cables … so pay attention as you work, measure your cable pattern gauge as needed, and do what you need to do to get the results you want!

CROCHET PATTERN: CABLED HEADBAND

By Erin - 11 Comments

This pattern is for an ultra versatile and classic looking cabled headband. Most people assume that making cables is difficult, but as soon as you master these two techniques:Front Post Double Crochet (fpdc) and Back Post Double Crochet (bpdc), cabling will be part of your everyday crocheting repertoire.

This pattern calls for sport weight yarn and the smaller E hook. If you would rather use a heavier weight yarn and a larger hook, have at it! With a DK or worsted yarn and a H hook this headband could also be used as a winter earwarmer!

Finished Size: 193 (48.26 cm) long and 2 1/43 (6.35 cm) wide

Note: the size of the headband may be adjusted using the instructions within the pattern.

Materials:approx. 40 yards Sportweight YarnCrochet Hook E (3.50 mm)

Gauge: not important in this pattern

Need help understanding the abbreviations and symbols? Check out the crochet abbreviation chart!

Crochet Pattern: Cabled Headband

Row 1: ch 12, dc in third ch from hook and in each ch across: 10 dc

Row 2: ch 2, turn, dc in first 2 dc, fpdc in next 2 dc, bpdc in next 2 dc, fpdc in next 2 dc, dc in last 2 dc: 4 dc, 4 fpdc, 2bpdc

Row 3: ch 2, turn, dc in first 2 dc, bpdc in next 2 fpdc, fpdc in next 2 bpdc, bpdc in next 2 fpdc, dc in last 2 dc: 4 dc, 4 bpdc, 2 fpdc

Row 4: ch 2, turn, dc in first 2 dc, skipping first bpdc, fpdc in next bpdc, fpdc in skipped bpdc, bpdc in next 2 dc, skipping next bpdc, fpdc in next bpdc, fpdc in skipped bpdc, dc in last 2 dc: 4 dc, 4 fpdc, 2 bpdc

Row 5: ch 2, turn, dc in first 2 dc, bpdc in next 2 fpdc, fpdc in next 2 bpdc, bpdc in next 2 fpdc, dc in last 2 dc: 4 dc, 4 bpdc, 2 fpdc

Row 6: ch 2, turn, dc in first 2 dc, skipping first bpdc, fpdc in next bpdc, fpdc in skipped bpdc, bpdc in next 2 dc, skipping next bpdc, fpdc in next bpdc, fpdc in skipped bpdc, dc in last 2 dc: 4 dc, 4 fpdc, 2 bpdc

Repeat Rows 5 and 6 until the length of the headband is 193 (48.26 cm) or when the length of the headband fits securely around the wearer's head.

Finishing: ch 2, turn, fold the headband in half with RS (in this case the RS in the side with the two twisting cables) held facing each other, join the two ends with an sc in each ch/dc. Finish off.

Cable needles are usually made of metal, plastic or wood. They can be straight, have a bend in the middle or be what's called a U-shape, which is really more like a J shape. Cable needles also come in different thicknesses to work with different weights of yarn.

To use a cable needle, simply slip the number of stitches required to work the cable onto the cable needle. Slip them as if you were purling them so that the stitches will not be twisted when you work them back into the knitting. Work the number of stitches indicated in the pattern and then work the stitches from the cable needle in the same order you would have worked them had they not been moved. Continue the row as normal.

Cable needles are very helpful when working cables, but they aren't required. If you don't want to buy a cable needle to try a cable knitting project, use a double-pointed needle of the same size as your knitting needles (so the stitches won't slide off easily).

Cable knitting is a style of knitting in which textures of crossing layers are achieved bypermuting stitches. For example, given four stitches appearing on the needle in the order ABCD, one might cross the first two (in front of or behind) the next two, so that in subsequent rows those stitches appear in the new order CDAB.

Cable knitting is a style of knitting in which textures of crossing layers are achieved by permuting stitches. For example, given four stitches appearing on the needle in the orderABCD, one might cross the first two (in front of or behind) the next two, so that in subsequent rows those stitches appear in the new order CDAB.

Methods

The stitches crossing behind are transferred to a small cable needle for storage while the stitches passing in front (or behind) are knitted. The former stitches are then transferred back to the original needle or knitted from the cable needle itself. Rather than use a cable needle, some knitters prefer to use a large safety pin or, for a single stitch, simply hold it in their fingers while knitting the other stitch(es). Cabling is typically done only when working on the right side of the fabric, i.e., every other row. This creates a spacer row, which helps the fabric to relax.

Cable knitting usually produces a fabric that is less flexible and more dense than typical knitting, having a much narrower gauge. This narrow gauge should be considered when changing from the cable stitch to another type of knitted fabric. If the number of stitches is not reduced, the second knitted fabric may flare out or pucker, due to its larger gauge.

Thus, ribbed cuffs on an aran sweater may not contract around the wrist or waist, as would normally be expected. Conversely, stitches may need to be added to maintain the gauge when changing from another knitted fabric such as stocking to a cable pattern.

Cables are usually done in stocking stitch and surrounded with reverse stocking; this causes the cables to stand out against a receding background. But any combination will do; for example, a background seed stitch in the regions bounded by cables often looks striking. Another visually intriguing effect is meta-cabling, where the cable itself is made up of cables, such as a three-cable plait made of strands that are themselves 2-cable plaits.

In such cases, the "inner" cables sometimes go their separate ways, forming beautiful, complex patterns such as the branches of a tree. Another interesting effect is to have one cable "pierce" another cable, rather than having it pass over or under the other.

Two cables should cross each other completely in a single row because making an intermediate crossing row of fewer stitches look good is very difficult. For example, where a pair of three-stitch-wide cables cross, all three stitches of one should cross over the three of the other cable.

Cable braids

Cables are often used to make braid patterns. Usually, the cables themselves are with a knit stitch while the background is done in purl. As the number of cables increases, the number of crossing patterns increases, as described by the braid group. Various visual effects are also possible by shifting the center lines of the undulating cables, or by changing the space between the cables, making them denser or more open.

A one-cable serpentine is simply a cable that moves sinusoidally left and right as it progresses. Higher-order braids are often made with such serpentines crossing over and under each other. A two-cable braid can look like a rope, if the cables always cross in the same way (e.g., left over right). Alternatively, it can look like two serpentines, one on top of the other.

A three-cable braid is usually a simple plait (as often seen styled in long hair), but can also be made to look like the links in a chain, or as three independent serpentines.

A four-cable braid allows for several crossing patterns. The five-cable braid is sometimes called the Celtic princess braid, and is visually interesting because one side is cresting while the other side is in a trough. Thus, it has a shimmering quality, similar to a kris dagger.

The six-cable braid is called a Saxon braid, and looks square and solid. This is a large motif, often used as a centerpiece of an aran sweater or along the neckline and hemlines.

The seven-cable braid is rarely used, possibly because it is very wide.It may be helpful to think of a cable pattern as a set of serpentine or wave-like cables, each one meandering around its own center line. A vast variety of cable patterns can be invented by changing the number of cables, the separations of their center lines, the amplitudes of their waves (i.e., how far they wander from their center line), the shape of the waves (e.g., sinusoidal versus triangular), and the relative position of the crests and troughs of each wave (e.g., is one wave cresting as another is crossing its center line?).

New cable patterns can also be inspired by pictures, scenes from nature, Celtic knotwork, and even the double helix of DNA.

Cable lattices

In some cases, one can form a lattice of cables, a kind of ribbing made of cables where the individual cable strands can be exchanged freely. A typical example is a set of parallel two-cable plaits in which, every so often, the two cables of each plait separate, going left and right and integrating themselves in the neighboring cables. In the process, the right-going cable of one plait crosses the left-going cable of its neighbor, forming an "X".

Cable textures

Many patterns made with cables do not have a rope-like quality. For example, a deep honeycomb pattern can be made by adjacent serpentines, first touching the neighbor on the left then the neighbor on the right. Other common patterns include a "Y"-like shape (and its inverse) and a horseshoe crab pattern.

Aran jumper

The Aran jumper is a style of jumper/sweater that takes its name from the Aran Islands off the west coast of Ireland. It is sometimes known as a fisherman's jumper. A classical fisherman's jumper is a bulky garment with prominent cable patterns on the chest, often cream-colored. The jumpers are distinguished by their use of complex textured stitch patterns, several of which are combined in the creation of a single garment. The word choice of "jumper" or "sweater" (or indeed other options such as "pullover" and "jersey") is largely determined by the regional version of English being spoken. In the case of Ireland and Britain, "jumper" is the standard word with "sweater" mainly found in tourist shops. The word used in Irish is geansaí, a gaelicization of guernsey which has been re-Anglicised to gansey in Hiberno-English.

Originally the jumpers were knitted using unscoured wool that retained its natural oils (lanolin) which made the garments water-resistant and meant they remained wearable even when wet. It was primarily the wives of island fishermen who knitted the jumpers.

Some stitch patterns have a traditional interpretation, often of religious significance. The honeycomb is a symbol of the hard-working bee. The cable, an integral part of the fisherman's daily life, is said to be a wish for safety and good luck when fishing. The diamond is a wish of success, wealth and treasure. The basket stitch represents the fisherman's basket, a hope for a plentiful catch.

Traditionally, an Aran jumper is made from undyed cream-coloured báinín (pronounced "bawneen"), a yarn made from sheep's wool, sometimes "black-sheep" wool. They were originally made with unwashed wool that still contained natural sheep lanolin, making it to an extent water-repellent. Up to the 1970s, the island women spun their own yarn on spinning wheels.

The jumper usually features four to six texture patterns each of which is about 5–10 cm (2–4 in) in width, that move down the jumper in columns from top to bottom. Usually, the patterns are symmetrical to a center axis extending down the center of the front and back panel. The patterns also usually extend down the sleeves as well.

History

There is debate about when island residents first started making the jumpers. Undoubtedly, residents of the islands produced a local version of a Gansey jumper similar to other areas of the Atlantic Isles for several centuries. Traditional Ganseys from neighbouring regions have much of the same cabling and pattern-work seen in Aran jumpers; however these Ganseys use different construction methods and are knit from a finer wool.

Some have suggested that the jumper is an ancient design that has been used on the islands for hundreds of years. Proponents of this theory often point to a picture in the Book of Kells that appears to depict an ancient "Aran jumper". Also, many megaliths around Europe depict similar patterns to those used in the knitting, which are carved into the stone, and date back several thousand years. However, it is more likely that the knitting stitches were modelled on these than that they evolved contemporaneously.

Most historians agree that far from being an ancient craft, Aran knitting was invented as recently as the early 1900s by a small group of enterprising island women, with the intention of creating garments not just for their families to wear but which could be sold as a source of income. These women adapted the traditional Gansey jumper by knitting with thicker wool and modifying the construction to decrease labor and increase productivity.

The first commercially available Aran knitting patterns were published in the 1940s by Patons of England. Vogue magazine carried articles on the garment in the 1950s, and jumper exports from the west of Ireland to the United States began in the early 1950s.

The development of the export trade during the 1950s and 1960s took place after P.A. Ó Síocháin organized an instructor, with the help of a grant from the Congested Districts Board for Ireland, to go to the islands and teach the knitters how to make garments to standard international sizings.

He commissioned the Irish artist Seán Keating, who had spent much time on the islands, to design and illustrate marketing brochures. Knitting became an important part of the islands' economy, and during the 1960s, even with all available knitters recruited from the three islands and from other parts of Ireland, he had difficulty in fulfilling orders from around the world.

Myths

Aran jumpers are often sold as a "fisherman sweater", suggesting that the jumper was traditionally used by the islands' famous fishermen. There is also some doubt about whether Aran jumpers were ever widely used by fishermen and many argue that the original jumpers with their untreated yarn would not have been suitable for this use. They are quite thick and stiff, which would probably restrict the movements of a fisherman.

However, the traditional Gansey jumpers which served as the model for the Aran jumpers have been worn by seamen and fishermen throughout the Atlantic Isles for centuries. Islanders can be seen wearing them in photographs taken early in the 20th century.

It is sometimes said that each fisherman (or his family) had a jumpers with a unique design, so that if he drowned and was found, maybe weeks later, on the beach, his body could be identified. This misconception may have originated with J.M. Synge's 1904 play Riders to the Sea, in which the body of a dead fisherman is identified by the hand-knitted stitches on one of his garments.

However, even in the play, there is no reference to any decorative or Aran-type pattern. The garment referred to is a plain stocking and it is identified by the number of stitches, the quote being "it's the second one of the third pair I knitted, and I put up three score stitches, and I dropped four of them". There is no record of any such event ever having taken place, nor is there any evidence to support there being a systematic tradition of family patterns.

There is, however, a long-standing tradition of jumper patterns having a regional or local identification. It is said that the county, or parish, or township of a sailor or a fisherman could be identified by his jumper pattern. Additionally, the wearer's initials were traditionally knit into the bottom of the garment, which would have been a far better indication of identity than the stitching pattern.

CLAN ARAN PATTERNS

The idea of clan Aran sweaters has widespread appeal to the Irish diaspora and is often used as a marketing tool. Most research concludes that there is no definitive evidence to support the concept of clan patterns, though it is possible that variations in design may have occurred in different families and regions. Anecdotal evidence varies on this point, with some older knitters saying that family patterns existed and others denying it.

Aran production today

While in the past, the majority of jumpers and other Aran garments were knitted by hand, today the majority of items for sale in Ireland and elsewhere are either machine knit or produced on a hand loom. There are very few people still knitting sweaters by hand on a commercial basis.

Machine-knitted jumpers tend to use finer wool and have less complex patterns, since many of the traditional stitches cannot be reproduced this way. They are the least expensive option.

Hand-looming allows more complicated stitches to be used, will have fewer stitches to the inch and be thicker. The best quality hand-loomed sweaters are almost indistinguishable from hand knit. Hand-knit sweaters tend to be more tightly knit, to have more complex stitch patterns and to be longer-lasting and they attract a significant price premium. By holding them up to light, the difference between the machine knit and hand knit is evident.\

The simple technique of cabling (crossing one group of stitches over another) lends itself to many interpretations in knitting. It's easy to do, you can make all kinds of interesting and imaginative cablepatterns. All it takes is a little patience and practice.

You can make any kind of cable by suspending (holding) a number of stitches on a cable needle (cn) while you knit that same number of stitches from the LH needle. Then you knit the suspended stitches either by returning them to the LH needle and knitting them or by knitting them straight from the cable needle. This process of knitting stitches out of order enables you to cross stitches, creating cables. Whether you're making simple or intricate cables, all you're doing is crossing stitches. Easy, right?

Standard or rope cables are the most basic cables. They generally cross stitches predictably up a single column of stitches. You can make a rope cable over almost any even number of stitches from two to twelve — or more. (To see how it's done, check out How to Knit Standard Cable Stitch [Rope Cable].)

- If you want to make a cable that looks like it's twisting to the left, you hold your suspended stitches in front of your work while you knit from the LH needle.
- If you want to make a cable that twists to the right, hold the suspended stitches in back.

Instructions for a 6-stitch left-twisting cable usually read something like this: Sl 3 sts to cn, hold infront, k3, k3 from cn. Instructions for that same cable, but twisting to the right, read like this: Sl 3 sts to cn, hold in back, k3, k3 from cn. (How to Read Knitting Abbreviations can help you decipher this explanation.)

You may also see abbreviations like C3F and C3B. The C before the number tells you that these stitches are cable stitches. The number tells you how many stitches are involved with this particular maneuver. The F or B indicates whether you should suspend the stitches to the front or the back of your work.

KNITTING A STANDARD CABLE

When you knit cables, you don't have to cross stitches on every row. (Thank goodness!) The row on which the stitches are crossed over each other is called the turning row. After the turning row, you work several plain rows, and then you work another turning row. Standard cables have the same number of plain rows between turning rows as there are stitches in the cable. If the cable is 6 stitches wide, for example, you work the turning row every 6 rows.

To try a 6-stitch left-twisting cable, cast on 14 stitches and then follow these steps:

1. Work Row 1 (RS): P4, k6, p4.

The first and last k4 (knit four) stitches make up the background fabric for your cable. The six stitches in the middle are where you'll form your cable stitches.

2. Work Row 2: K4, p6, k4.

3. Work Row 3: P4, k6, p4. (same as row 1).

4. Work Row 4: k4, p6, k4 (same as row 2).

5. Work Row 5, the turning row: P4, sl 3 sts to cn and hold in front (see Figure 1a), k3 from LH needle, k3 from cn (see Figure 1c), p4.

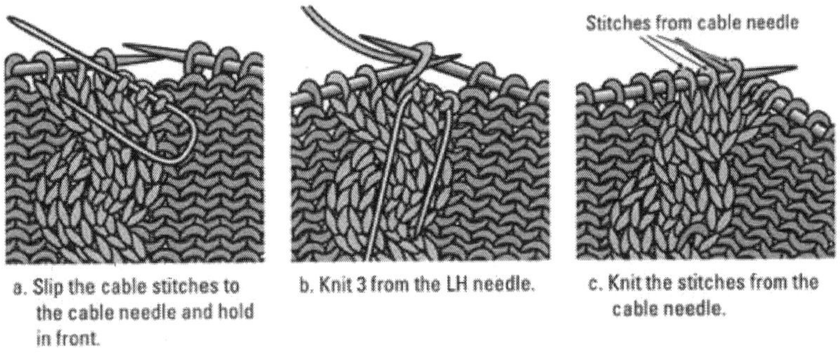

Fig 1: Working the turning row.

6. Work Row 6: K4, p6, k4 (same as Row 2).

Instead of knitting stitches directly from the cable needle, you may prefer to return the suspended stitches to the LH needle before you knit them. Try both ways and use whichever technique is more comfortable to you.

When you suspend stitches on the cable needle, let the cable needle dangle down in front of your work, giving the yarn a slight tug to keep it taut (you don't need to close the gap).

7. Repeat Steps 1 through 6 and watch your stockinette stitches become a cabled rope.

When you work cables, you go back and forth from purl stitches to knit stitches. From a knit to a purl stitch, bring your yarn between the needles to the front before you make the next stitch. From a purl to a knit stitch, bring your yarn between the needles to the back before you make the next stitch.

CHECKING GAUGE IN A CABLE PATTERN

The combination of knit panels with purl panels (think ribs) and crossing stitches over stitches causes cable patterns to pull in widthwise. A sweater worked in a cable pattern will be significantly narrower than one worked in the same number of stitches in stockinette stitch. You'll need more yarn and more stitches for a cable sweater than for one of the same dimensions in a knit/purl pattern.

If you decide to add a cable (or several) to a plain sweater, be sure to increase enough stitches after you knit your border to maintain the overall width. Although there are no hard and fast rules, you'll be safe if you add 1 to 2 for every 4 stitches in your cable. If you have a ribbed border, you can add the stitches evenly on the last ribbed row.

If you're making a project in a repeating cable pattern, be sure to work a large enough swatch to be able to measure gauge accurately. (See How to Knit a Gauge Swatch.) The swatch should include at least two repeats of the cable pattern horizontally and vertically. If you're working several different cables, you have to check your gauge over each one.

Feel like you're wasting time making swatches to check gauge? Make two gauge swatches and sewthem together for a cabled pillow or bag.

Starting and stopping

Cables should begin and end on a row between turnings where the cable stitches are relaxed and the fabric is smooth. Because the cable stitches spread out a bit between turning rows, for the smoothest transition between the beginnings and endings of cables, plan to increase a stitch or two over the cable stitches when beginning a cable and decrease a stitch or two when binding off over a cable.

To start a cable pattern

If you're adding a cable(s) to an uncabled sweater pattern, for every 4 stitches in the planned cable(s), add 1 or 2 stitches to the number of stitches to cast on. (For step-by-step instructions for casting on, see How to Use the Cable Cast-On Method.) Then work a few rows in the knit/purl pattern you've established for your cables (which How to Knit the Stockinette Stitch describes) before working a turning row.

To end a cable pattern

When you're binding off cables, decrease across the top of the cable 1 or 2 stitches for every 4 stitches in the cable to prevent the bound-off edge from flaring out. Follow these steps for a neat finish:

1. Bind off the background stitches as they present themselves (knit the knits and purl the purls as you bind off) until you reach the cabled stitches.

2. When you get to the cable stitches, bind off as follows: *k1, bind off, k2tog, bind off, k1, bind off; rep from * to the end of the cable (stockinette) stitches.

3. After binding off the cable stitches, go back to binding off the stitches of the background as they present themselves.

8

FINGER KNITTING TECHNIQUE

An easy and fun craft, finger-knitting makes for a great way to busy the hands while quieting the mind. Contributor Katie Gast explains the many applications there are for the finished product—from necklaces to give as gifts to garlands for a tree or mantel. Finger knitting is a fun, productive way to spend your free time when you need something to do. When you're finished, you'll have a beautiful rope of yarn that you can use in many different ways, like a key ring, hair decoration, belt, or bag handle.

No needles are involved in this technique and it can be easily mastered by young and old alike-Plus it is safe to do on planes. Finger-knitting is done entirely by hand, using a simple yarn-looping technique! One you learn it you won't know when to stop. You can finger knit a flower or take a number of separate pieces and sew them together for a coaster.

You can even use some of the bulky yarn to make a lovely scarf. What, might you ask, is finger knitting? Why simply knitting with one's own fingers in place of the traditional knitting needles. Whether you are searching for a new fiber experience, or stuck in a needle-free-zone, finger knitting is for you. The process is amazingly simple, and the only supplies required are some yarn of your choosing and your own fingers.

Finger knitting is a great, kid-friendly project (read: cheap and easy) that anyone can do. Follow my fingers and you'll be knitting right along in just a snap! Finger knitting produces a long, thin strip of stockinette stitch. The resulting thin strip may not seem like much, but think of the possibilities...scarves with single or double strips, three strips braided together, handles, straps, ties, sew many together for a throw blanket, all without needles. This is the ultimate on-the-go, take-it-anywhere, super-simple knitting project. For this project, you can use absolutely any type (and mixture) of yarns imaginable. I recommend using a bulky weight yarn or working with two or three lighter weight yarns together at the same time. Play with novelty yarns, different textures, or crazy color combos. Finger knitting is a great way to experiment with putting different yarns together without a serious commitment. Mix it up and have fun! I chose to work with two wools held together — both from Kraemer Yarns, Naturally Nazareth Aran in Summer and Mauch Chunky in Kiwi.

USES OF FINGER KNITTING

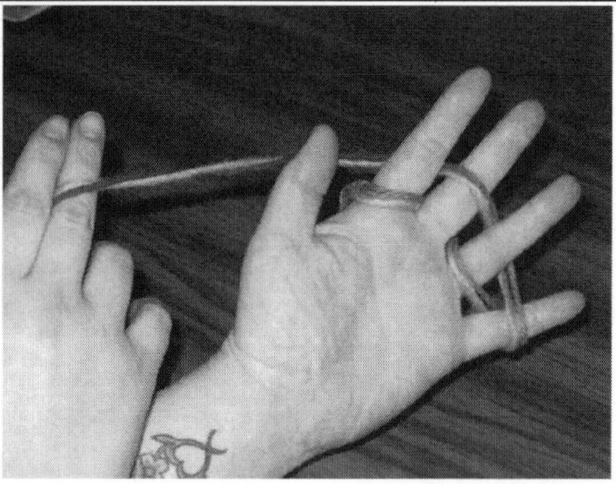

Begin with your left hand facing you (assuming you are right handed. For lefties, begin with the right hand facing you). The tail of your yarn(s) should be placed in the area between your left thumb and index finger. Begin casting on by wrapping the working yarn over the index finger, behind the middle finger, over the ring finger, etc.

Once you wrap around the pinkie, continue back the other direction until your reach the index finger once again. Go around the index finger a second time and wrap all fingers in the same manner slightly above the wraps from the first pass. Try to maintain loose wraps. You not only need to be able to work the stitches, you also don't want your lovely fingers to turn blue.

The First Row

Beginning at the pinkie, lift the wrapped yarn on the bottom (nearest your palm) up and over the tip of the finger, keeping the top wrap on the finger as you pass the lower wrap over and off.Remember that in these photos, I'm using two strands held together as one. Continue this process along the hand until your reach the index finger. Be careful as you lift the stitch on the index finger since the bottom wrap is actually the tail of your yarn. (Just don't let it get away from you!)

All Following Rows

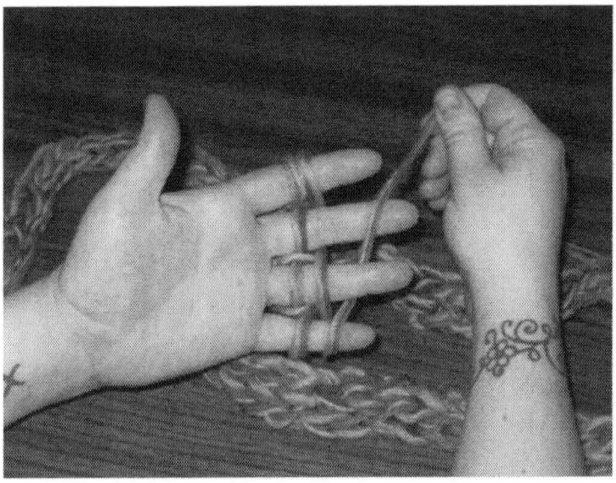

For the next row, and every other row, begin by once again wrapping the working yarn around all four fingers as in step one. Next, each lower stitch is passed over the upper stitch and lifted off the finger as in step two. This two-step process continues until the work is the desired length. When making scarves, I like to knit about eight to ten feet of finger knitting and double-up the strips around the neck. Should you need to put down your finger knitting sometime during construction, I recommend sliding the working stitches onto a ballpoint pen and hooking them under the paper-clippie thing at the top. (This is a needle-free area, remember?) When ready to begin again, place the working stitches back onto the fingers with the stockinette side of the work facing the back of your hand.

Binding Off

Once the chosen length is reached, the knitter must bind off the working stitches. On this last row, do not wrap the fingers. Each finger should have only one loop on it. Lift off the loop on the pinkie finger and place it onto the ring finger. On the ring finger, lift the bottom loop up and over the top loop and off the finger.

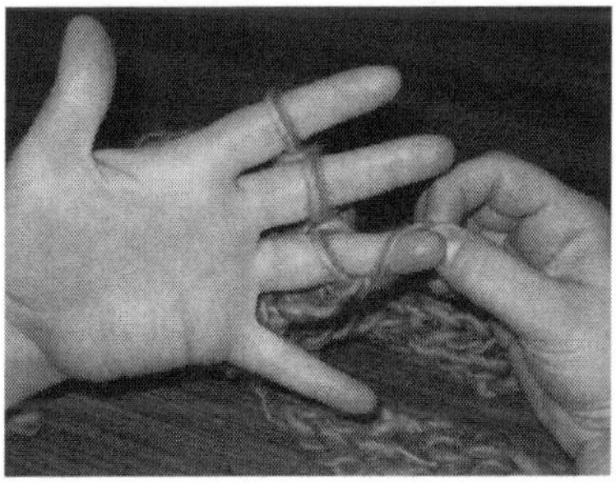

Next, place the remaining loop from the ring finger onto the middle finger and repeat the lifting off/moving over step until one loop remains on the index finger. To finish, simply cut a tail and pass it through the remaining loop, pulling it tight.

Steps

Loop the yarn around your thumb of your non-dominant hand. Don't tie a knot. The tail should be on the back of your hand, and you should press the side of your thumb against your index finger to hold it tightly.

Weave the yarn over your fingers four times. Take the other, "working" end of the yarn and pull it over your index finger, keeping the tail tightly gripped between the side of your thumb and index finger. Go under your middle finger, over your ring finger, and under your pinky. Loop around, going over the pinky, under the ring finger, over the middle finger, and under the index finger. Do this sequence one more time so that each finger has two complete loops on it.

Pull up the bottom loops. Take the bottom "line" of yarn on your pinky and pull it over your finger, taking it up above the first line. That loop of yarn should now be on the back of your hand. Repeat with all your fingers, moving from pinky to middle finger.

- When you reach your index finger, simply move the tail that was between your thumb and index finger between your index and middle fingers. It should be hanging down the back of your hand. This counts as the first "bottom loop" for your index finger.

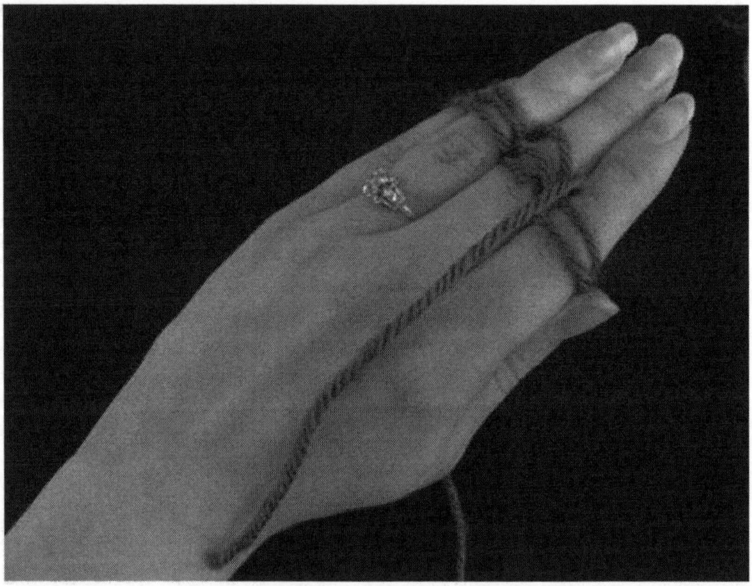

Weave the yarn in and out of your fingers again until there is another line. Pull the bottom loops up again. Continue for as long as you want or until you run out of yarn. A nice rope-like structure of yarn should form at the back of your hand.

Finish off. When your rope is as long as you want it, stop when you have one loop on each finger.

- Take the loop from your pinky, and put it on your ring finger. Pull the bottom loop up from your ring finger and over to the back of your hand.
- Move the loop from your ring finger to your middle finger. Again, pull up the bottom loop over to the back of your hand.
- Move the loop from your middle finger to your index finger. Pull up the bottom loop again. You should be left with only one loop on your index finger now.
- Slip the loop off your index finger.

Cut your working yarn a few inches away from the loop, and pull the tail through the loop. Tug a few times to tighten. You can also pull on the original tail to tighten it as well.

Tie the ends of the rope in a square knot to secure.

Tips

- You can tighten the rope every few stitches by pulling on the end of the rope.
- It is much better to do this all at the one time, or you might lose the rhythm and forget where you were.

- You can place the loops on your fingers onto a pencil or straw to hold your place so you can come back to your weaving another time.
- You can make scarves, belts, bracelets, headbands, pot holders, and more.
- If you want to make a thinner, faster strand, simply do the above method using only one finger instead of four.

Warnings

- If you see loose loops in your rope, don't cut it. If the loop is close to your knuckles, just pull out the stitches back to the problem. Then redo that part.
- If you suffer from Repetitive Stress Injury, remember to take frequent breaks.

Finger knitting is a great way to keep little hands busy, while creating long cords in a flash! Kids dig it because it's faster and less cumbersome than traditional knitting. Once knitted, the cords can be made into things like googly-eyed snakes, purse handles, gift ribbons, or my son Tanner's favorite project du jour, the Twisty Scarf, which we'll show you how to knit. Hey, don't just sit there, let your fingers do the knitting!

Materials

Yarn

Your hands

Directions

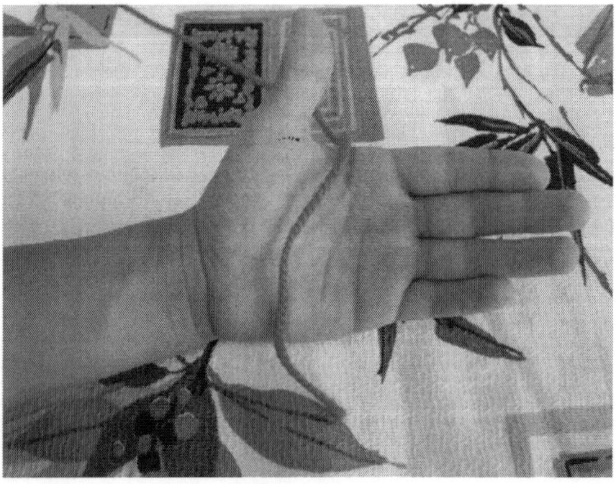

Step 1: Leaving about a 6" tail, lay the yarn across your left-hand palm (reverse, if you're a lefty).

Step 2: Using your right hand, wrap yarn clockwise around your left-hand index finger.

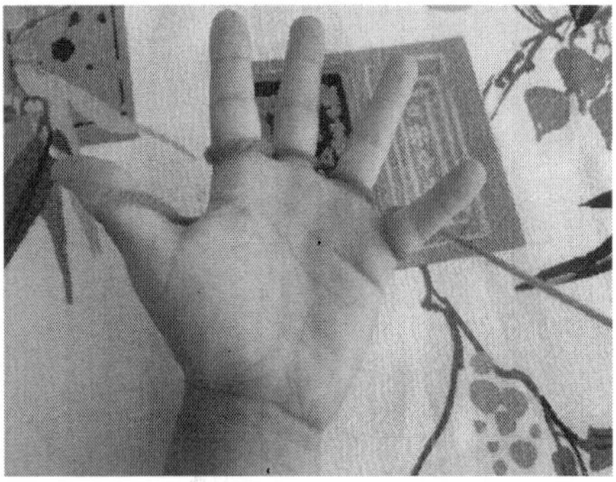

Steps 3: Continuing as in Step 2, wrap the yarn clockwise around your middle, ring, and pinky fingers. You'll now have loops on all fingers, excluding the thumb.

Step 4: Lay the yarn across your fingers, above the loops.

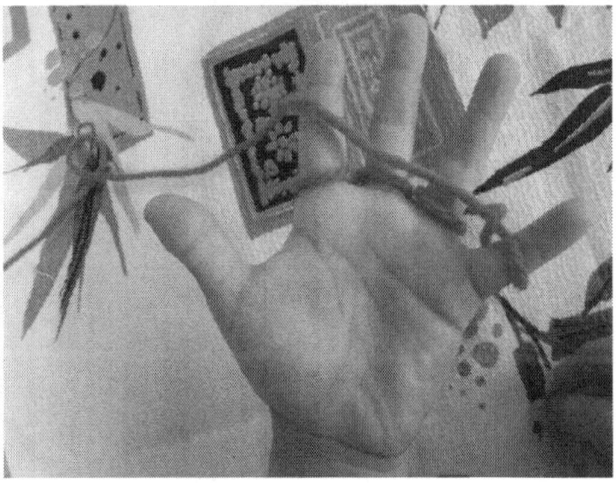

Step 5: Lift your pinky finger loop over the yarn, and let it drop off your finger.

Steps 6: Continuing as in Step 5, lift the loop over the yarn and let it drop off on the ring, middle, and pointer fingers. You've just completed your first row!

Steps 7: Repeat Steps 4–6 until your finger-knitted cord is the desired length.

Keep in mind that at the end of each row, the yarn end will be on the opposite side of the palm. Simply lay the yarn across your palm, beginning from whichever direction the yarn is oriented.

To finish off, cut the yarn, leaving a 6" tail. Weave the tail through loops, letting them fall off your fingers. Pull tight and knot. That's all there is to it!

Twisty Scarf

Materials:

3 different colors of the same weight yarn

Tapestry needle (optional)

Tape

Directions

Step 1: Finger knit one 48" cord in each of the three colors.

Step 2: Tape the ends of all three cords down to a table or the floor.

Step 3: Braid the cords together.

Step 4: Using a tapestry needle and a strand of yarn, sew an "X" at either end of the braid to tack the cord pieces together. Alternatively, you can skip the sewing and just knot the ends.

Step 5: Trim the yarn tails to about 1"–2".

In the olden days, finger knitting was primarily used to introduce the young children, especially girls, to the intricacies of knitting. It was used to teach young children how to carry out the process of knitting though the difference here is that no needles were used. But with the passage of time, the applications of finger knitting became quite diverse.

One of the areas through which finger knitting has become very important is as a tool for educating young children. Very little in terms of resources are required when using finger knitting to teach children, only fingers and yarns are necessary. A top child psychiatrist has stated that brains are known to discover what hands explore and this has become the motivation for teachers using finger knitting as a teaching tool.

The import of what the psychiatrist above said was that through finger knitting, children learn how to coordinate the different activities their fingers carry out with their brains. As fingers create loops and determine the level of tension in the yarn children explore the full extent of what their tiny brains are capable of. Through finger knitting, these young children learn how to solve problems.

Finger knitting helps young children develop their creative, intellectual as well as physical sides thus becoming all round students. In learning to come up with beautifully crafted designs and loops, their brains develop faster in terms of developing better coordination with the fingers and other organs of the body. Consequently, this attribute reassures many on the importance of finger knitting in developing young minds quickly.

Finger blindness is a condition that bedevils people who never got the chance to develop the use of their fingers better while at a young age. A child unable to develop the proper use of his fingers or to coordinate or even differentiate the true abilities and purposes of their fingers is said to be suffering from finger blindness. Finger blindness can hinder the full development and growth of the affected children. Differentiation or proper appreciation of different colors is another attribute that finger knitting helps inculcate in young children. Teachers who use different colored yarns provide a faster avenue of learning for the children under their tutelage. This is more evident when teachers use brightly colored yarns as these help hold the attention of the young children for longer spans of time than would normally have been possible.

Finger knitting teaches children on the important aspect of balancing. When the yarn is pulled toot tightly across the fingers the stitches become too tight thus hard to work with. On the other hand, when the yarn is left too loosely, it forms very big loops. Consequently, depending on what they desire to design, children can be taught just how to maintain the right levels of tension and balance through finger knitting.

Finger knitting presents tremendous opportunities for young children to learn how to use their time well while designing some very wonderful items. Children learn some very important lessons on how creative they can be. Finger knitting has proved to be a very popular teaching method for children while having fun in the process of learning.

Finger knitting is a great past-time to pick up for both adults and children alike. If you are looking for a great pastime to start or you want to find ways on how to earn some cash from a pasttime that you love.

Crochet

Crochet is a process of creating fabric from yarn,thread, or other material strands using a crochet hook. The word is derived from the French word "crochet", meaning hook. Hooks can be made of materials such as metals, woods or plastic and are commercially manufactured as well as produced by artisans.

Crocheting, like knitting, consists of pulling loops through other loops, but additionally incorporates wrapping the working material around the hook one or more times. Crochet differs from knitting in that only one stitch is active at one time (exceptions being Tunisian crochet and Broomstick lace), stitches made with the same diameter of yarn are comparably taller, and a single crochet hook is used instead of two knitting needles. Additionally, crochet has its own system of symbols to represent stitch types.

Origins

Lis Paludan theorizes that crochet evolved from traditional practices in Arabia, South America, or China, but there is no decisive evidence of the craft being performed before its popularity in Europe during the 19th century. The earliest written reference to crochet refers to shepherd's knitting from The Memoirs of a Highland Lady by Elizabeth Grant (1797–1830) in the 19th century. The first published crochet patterns appeared in the Dutch magazine Pénélopé in 1824.

Other indicators that crochet was new in the 19th century include the 1847 publication A Winter's Gift, which provides detailed instructions for performing crochet stitches, although it presumes that readers understand the basics of other needlecrafts. Early references to the craft in Godey's Lady's Book in 1846 and 1847 refer to crotchet before the spelling standardized in 1848.

Kooler proposes that early industrialization is key to the development of crochet. Machine spun cotton thread became widely available and inexpensive in Europe and North America after the invention of the cotton gin and the spinning jenny, displacing hand spun linen for many uses. Crochet technique consumes more thread than comparable textile production methods and cotton is well suited to crochet.

Knit and knotted textiles survive from very early periods, but there are no surviving samples of crocheted fabric in any ethnological collection, or archeological source prior to 1800. These writers point to the tambour hooks used in tambour embroidery in France in the 18th century, and contend that the hooking of loops through fine fabric in tambour work evolved into "crochet in the air." Most samples of early work claimed to be crochet turn out to actually be samples of nålebinding.

Donna Kooler identifies a possible problem with the tambour hypothesis: period tambour hooks that survive in modern collections cannot produce crochet because the integral wing nut necessary for tambour work interferes with attempts at crochet.

However, Mrs. Gaugain, in her 1840 The Lady's Assistant for Executing Useful and Fancy Designs in Knitting, Netting, and Crotchet Work, refers to "Tambour, or Crotchet," then proceeds to call it "tambour" in all the instructions, indicating a strong connection believed in at the time of crochet's beginning, and that it was, perhaps, the older name. That those hooks that survive cannot be used is the constant problem in archaeology: what was commonly used may have been usually worn out and didn't survive. Tambour lace on fine net is commonly taught with the crochet hook, these hooks with "integral wing nuts" being an expensive item.

Early crochet hooks ranged from primitive bent needles in a cork handle, used by poor Irish lace workers, to expensively crafted silver, brass, steel, ivory and bone hooks set into a variety of handles, some of which were better designed to show off a lady's hands than they were to work with thread. By the early 1840s, instructions for crochet were being published in England, particularly by Eleanor Riego de la

Blanchardiere and Frances Lambert. These early patterns called for cotton and linen thread for lace, and wool yarn for clothing, often in vivid color combinations.

In the 19th century, as Ireland was facing the Great Irish Famine (1845-1849), crochet lacework was introduced as a form of famine relief (the production of crocheted lace being an alternative way of making money for impoverished Irish workers). Mademoiselle Riego de la Blanchardiere is generally credited with the invention of Irish Crochet, publishing the first book of patterns in 1846. Irish lace became popular in Europe and America, and was made in quantity until the first World War.

MODERN PRACTICE AND CULTURE

Fashions in crochet changed with the end of the Victorian era in the 1890s. Crocheted laces in the new Edwardian era, peaking between 1910 and 1920, became even more elaborate in texture and complicated stitching.

The strong Victorian colours disappeared, though, and new publications called for white or pale threads, except for fancy purses, which were often crocheted of brightly colored silk and elaborately beaded. After World War I, far fewer crochet patterns were published, and most of them were simplified versions of the early 20th century patterns.

After World War II, from the late 40s until the early 60s, there was a resurgence in interest in home crafts, particularly in the United States, with many new and imaginative crochet designs published for colorful doilies, potholders, and other home items, along with updates of earlier publications.

These patterns called for thicker threads and yarns than in earlier patterns and included wonderful variegated colors. The craft remained primarily a homemaker's art until the late 1960s and early 1970s, when the new generation picked up on crochet and popularized granny squares, a motif worked in the round and incorporating bright colors.

Although crochet underwent a subsequent decline in popularity, the early 21st century has seen a revival of interest in handcrafts and DIY, as well as great strides in improvement of the quality and varieties of yarn. There are many more new pattern books with modern patterns being printed, and most yarn stores now offer crochet lessons in addition to the traditional knitting lessons.

There are many books you can purchase from local book stores to teach yourself how to crochet whether it be as a beginner or more intermediate, there are also many books for children and teenagers who are hoping to take up the hobby. One of the books available for teens is 'The Cool Girl's Guide to Crochet' by Nicki

Trench. Filet crochet, Tunisian crochet, tapestry crochet,broomstick lace, hairpin lace, cro-hooking, and Irish crochet are all variants of the basic crochet method

Crochet has experienced a revival on the catwalk. Christopher Kane's Fall 2011 Ready-to-Wear collection makes intensive use of the granny square, one of the most basic of crochet motifs. Crochet has been utilized many a time by designers on the popular reality show Project Runway. Additionally, websites such as Etsy and Ravelry have made it easier for individual hobbyists to sell and distribute their patterns or projects across the internet.

Materials

Basic materials required for crochet are a hook and some type of material that will be crocheted, most commonly yarn or thread. Additional tools are convenient for keeping stitches counted, measuring crocheted fabric, or making related accessories. Examples include cardboard cutouts, which can be used to make assels, fringe, and many other items; a pom-pom circle, used to make pom-poms; a tape measure and a gauge measure, both used for measuring crocheted work and counting stitches; a row counter; and occasionally plastic rings, which are used for special projects.

Hook

The crochet hook comes in many sizes and materials, such as bone, bamboo, aluminum, plastic and steel. Sizing is categorized by the diameter of the hook's shaft, and a crafter aims to create stitches of a certain size in order to reach a particular gauge specified in a given pattern. If gauge is not reached with one hook, another is used until the stitches made are the needed size.

Crafters may have a preference for one type of hook material over another due to aesthetic appeal, yarn glide, or hand disorders such as arthritis, where bamboo or wood hooks are favored over metal for the perceived warmth and flexibility during use. Hook grips and ergonomic hook handles are also available to assist crafters.

Steel crochet hooks range in size from 0.4 to 3.5 millimeters, or from 00 to 16 in American sizing. These hooks are used for fine crochet work such as doilies and lace.

Aluminum, bamboo, and plastic crochet hooks are available from 2.5 to 19 millimeters in size, or from B to S in American sizing.

Artisan-made hooks are often made of hand-turned woods, sometimes decorated with semi-precious stones or beads.

Crochet hooks used for Tunisian crochet are elongated and have a stopper at the end of the handle, while double-ended crochet hooks have a hook on both ends of the handle. There is also a double hooked apparatus called a Cro-hook that has become popular.

A hairpin loom is often used to create lacey and long stitches, known as hairpin lace. While this is not in itself a hook, it is a device used in conjunction with a crochet hook to produce stitches.

The patent-pending Sharp Crochet Hook (www.sharpcroch ethook.com) is used to crochet edgings through fabric because the sharp-pointed head glides through fabric easier than normal crochet hooks. It can be used to crochet edgings on baby blankets, skirt hems, gloves, socks, pillow cases, towels, etc.

Yarn

Yarn for crochet is usually sold as balls or skeins (hanks), although it may also be wound on spools or cones. Skeins and balls are generally sold with a yarn band, a label that describes the yarn's weight, length, dye lot, fiber content, washing instructions, suggested needle size, likely gauge, etc. It is a common practice to save the yarn band for future reference, especially if additional skeins must be purchased. Crocheters generally ensure that the yarn for a project comes from a single dye lot.

The dye lot specifies a group of skeins that were dyed together and thus have precisely the same color; skeins from different dye lots, even if very similar in color, are usually slightly different and may produce a visible stripe when added onto existing work. If insufficient yarn of a single dye lot is bought to complete a project, additional skeins of the same dye lot can sometimes be obtained from other yarn stores or online.

The thickness or weight of the yarn is a significant factor in determining the gauge, i.e., how many stitches and rows are required to cover a given area for a given stitch pattern. Thicker yarns generally require large-diameter crochet hooks, whereas thinner yarns may be crocheted with thick or thin hooks. Hence, thicker yarns generally require fewer stitches, and therefore less time, to work up a given project.

Patterns and motifs are coarser with thicker yarns and produce bold visual effects, whereas thinner yarns are best for refined or delicate patternwork. Yarns are standardly grouped by thickness into six categories: superfine, fine, light, medium, bulky and superbulky. Quantitatively, thickness is measured by the number of wraps per inch (WPI). The related weight per unit length is usually measured in tex or dernier.

Before use, hanks are wound into balls in which the yarn emerges from the center, making crocheting easier by preventing the yarn from becoming easily tangled. The winding process may be done by hand, or with a ballwinder and swift.

A yarn's usefulness is judged by several factors, such as its loft (its ability to trap air), itsresilience (elasticity under tension), its washability and colorfastness, its hand (its feel, particularly softness vs. scratchiness), its durability against abrasion, its resistance topilling, its hairiness (fuzziness), its tendency to twist or untwist, its overall weight and drape, its blocking and felting qualities, its comfort (breathability, moisture absorption, wicking properties) and its appearance, which includes its color, sheen, smoothness and ornamental features.

Other factors include allergenicity, speed of drying, resistance to chemicals, moths, and mildew, melting point and flammability, retention of static electricity, and the propensity to accept dyes. Desirable properties may vary for different projects, so there is no one "best" yarn.

Although crochet may be done with ribbons, metal wire or more exotic filaments, most yarns are made by spinning fibers. In spinning, the fibers are twisted so that the yarn resists breaking under tension; the twisting may be done in either direction, resulting in an Z-twist or S-twist yarn.

If the fibers are first aligned by combing them and then the spinner uses a worsted type drafting method such as the short forward draw, the yarn is smoother and called aworsted; by contrast, if the fibers are carded but not combed and the spinner uses a woolen drafting method such as the long backward draw, the yarn is fuzzier and called woolen-spun. The fibers making up a yarn may be continuous filament fibers such as silk and many synthetics, or they may be staples (fibers of an average length, typically a few inches); naturally filament fibers are sometimes cut up into staples before spinning.

The strength of the spun yarn against breaking is determined by the amount of twist, the length of the fibers and the thickness of the yarn. In general, yarns become stronger with more twist (also called worst), longer fibers and thicker yarns (more fibers); for example, thinner yarns require more twist than do thicker yarns to resist breaking under tension. The thickness of the yarn may vary along its length; a slub is a much thicker section in which a mass of fibers is incorporated into the yarn.

The spun fibers are generally divided into animal fibers, plant and synthetic fibers. These fiber types are chemically different, corresponding to proteins, carbohydrates and synthetic polymers, respectively. Animal fibers include silk, but generally are long hairs of animals such as sheep (wool), goat (angora, or cashmere goat), rabbit (angora), llama, alpaca, dog, cat, camel, yak, and muskox(qiviut). Plants used for fibers include cotton, flax (for linen), bamboo, ramie, hemp, jute, nettle, raffia, yucca, coconut husk, banana trees, soy and corn. Rayon and acetate fibers are also produced from cellulose mainly derived from trees. Common synthetic fibers include acrylics, polyesters such as dacron and ingeo, nylon and other polyamides, and olefins such as poly-propylene.

Of these types, wool is generally favored for crochet, chiefly owing to its superior elasticity, warmth and (sometimes) felting; however, wool is generally less convenient to clean and some people are allergic to it. It is also common to blend different fibers in the yarn, e.g., 85% alpaca and 15% silk. Even within a type of fiber, there can be great variety in the length and thickness of the fibers; for example, Merinowool and Egyptian cotton are favored because they produce exceptionally long, thin (fine) fibers for their type.

A single spun yarn may be crochet as is, or braided or plied with another. In plying, two or more yarns are spun together, almost always in the opposite sense from which they were spun individually; for example, two Z-twist yarns are usually plied with an S-twist. The opposing twist relieves some of the yarns' tendency to curl up and produces a thicker, balanced yarn.

Plied yarns may themselves be plied together, producing cabled yarns or multi-stranded yarns. Sometimes, the yarns being plied are fed at different rates, so that one yarn loops around the other, as in bouclé. The single yarns may be dyed separately before plying, or afterwords to give the yarn a uniform look.

The dyeing of yarns is a complex art. Yarns need not be dyed; or they may be dyed one color, or a great variety of colors. Dyeing may be done industrially, by hand or even hand-painted onto the yarn. A great variety of synthetic dyes have been developed since the synthesis of indigo dye in the mid-19th century; however, natural dyes are also possible, although they are generally less brilliant. The color-scheme of a yarn is sometimes called its colorway. Variegated yarns can produce interesting visual effects, such as diagonal stripes.

Process

Crocheted fabric is begun by placing a slip-knot loop on the hook (though other methods, such as a magic ring or simple folding over of the yarn may be used), pulling another loop through the first loop, and repeating this process to create a chain of a suitable length. The chain is either turned and worked in rows, or joined to the beginning of the row with a slip stitch and worked in rounds.

Rounds can also be created by working many stitches into a single loop. Stitches are made by pulling one or more loops through each loop of the chain. At any one time at the end of a stitch, there is only one loop left on the hook. Tunisian crochet, however, draws all of the loops for an entire row onto a long hook before working them off one at a time. Like knitting, crochet can be worked either flat or in the round.

International crochet terms and notations

In the English-speaking crochet world, basic stitches have different names that vary by country. The differences are usually referred to as UK/US or British/ American. To help counter confusion when reading patterns, a diagramming system using a standard international notation has come into use (illustration, left).

Another terminological difference is known as tension (UK) and gauge (US). Individual crocheters work yarn with a loose or a tight hold and, if unmeasured, these differences can lead to significant size changes in finished garments that have the same number of stitches. In order to control for this inconsistency, printed crochet instructions include a standard for the number of stitches across a standard swatch of fabric. An individual crocheter begins work by producing a test swatch and compensating for any discrepancy by changing to a smaller or larger hook. North Americans call this gauge, referring to the end result of these adjustments; British crocheters speak of tension, which refers to the crafter's grip on the yarn while producing stitches.

Differences from and similarities to knitting

One of the more obvious differences is that crochet uses one hook while much knitting uses two needles. In most crochet, the artisan usually has only one live stitch on the hook (with the exception being Tunisian crochet), while a knitter keeps an entire row of stitches active simultaneously. Dropped stitches, which can unravel a fabric, rarely interfere with crochet work, due to a second structural difference between knitting and crochet.

In knitting, each stitch is supported by the corresponding stitch in the row above and it supports the corresponding stitch in the row below, whereas crochet stitches are only supported by and support the stitches on either side of it. If a stitch in a finished crocheted item breaks, the stitches above and below remain intact, and because of the complex looping of each stitch, the stitches on either side are unlikely to come loose unless heavily stressed.

Round or cylindrical patterns are simple to produce with a regular crochet hook, but cylindrical knitting requires either a set of circular needles or three to five special double-ended needles. Many crocheted items are composed of individual motifs which are then joined together, are by sewing or crocheting, whereas knitting is usually composed of one fabric, such as entrelac.

Freeform crochet is a technique that can create interesting shapes in three dimensions because new stitches can be made independently of previous stitches almost anywhere in the crocheted piece. It is generally accomplished by building shapes or structural elements onto existing crocheted fabric at any place the crafter desires.

Knitting can be accomplished by machine, while many crochet stitches can only be crafted by hand. The height of knitted and crocheted stitches is also different: a single crochet stitch is twice the height of a knit stitch in the same yarn size and comparable diameter tools, and a double crochet stitch is about four times the height of a knit stitch.

While most crochet is made with a hook, there is also a method of crocheting with a knitting loom. This is called loomchet. Slip stitch crochet is very similar to knitting. Each stitch in slip stitch crochet is formed the same way as a knit or purl stitch which is then bound off. A person working in slip stitch crochet can follow a knitted pattern with knits, purls, and cables, and get a similar result.

It is a common perception that crochet produces a thicker fabric than knitting, tends to have less "give" than knitted fabric, and uses approximately a third more yarn for a comparable project than knitted items. Though this is true when comparing a single crochet swatch with a stockinette swatch, both made with the same size yarn and needle/hook, it is not necessarily true for crochet in general. Most crochet uses far less than 1/3 more yarn than knitting for comparable pieces, and a crocheter can get similar feel and drape to knitting by using a larger hook or thinner yarn.

Tunisian crochet and slip stitch crochet can in some cases use less yarn than knitting for comparable pieces. According to sources claiming to have tested the 1/3 more yarn assertion, a single crochet stitch (sc) uses approximately the same amount of yarn as a knit stitch. All crochet stitches which require a yarn over to complete the stitch operation will use slightly more yarn. Cluster stitches, which are in fact multiple stitches worked together,will use the most yardage.

Charity

It has been very common for people and groups to crochet clothing and other garments and then donate them to soldiers during war. People have also crocheted clothing and then donated it to hospitals, for sick patients and also for newborn babies. Sometimes groups will crochet for a specific charity purpose, such as crocheting for homeless shelters, nursing homes, etc. It is also becoming increasingly popular to crochet hats (commonly referred to as "chemo caps") and donate them to cancer treatment centers, for those undergoing chemotherapy.

Crochet patterns have an underlying mathematical structure—the pattern created by the regular presence or omission of stitches is the very essence of this artform. The similarities to Base2 math, with its series of 0s and 1s, are obvious. That is to say, a present stitch is like a "1", and a missing stitch is like a "0". The craft has been used to illustrate shapes in hyperbolic space that are difficult to reproduce using other media or are difficult to understand when viewed two-dimensionally. A hyperbolic model of a coral reef has also been constructed for environmental purposes

A paper model based on the pseudosphere was created by William Thurston, however, it was quite delicate. Crochet has been used by mathematician Daina Taimina in order to create a version of the hyperbolic plane. Daina Taimina used the art of crochet to create a strong, durable model (see related image), which received an exhibition by the Institute For Figuring.

Architecture

In "Style in the technical arts", Gottfried Semper looks at the textile with great promise and historical precedent. In Section 53, he writes of the "loop stitch, or Noeud Coulant: a knot that, if untied, causes the whole system to unravel." In the same section, Semper confesses his ignorance of the subject of crochet but believes strongly that it is a technique of great value as a textile technique and possibly something more.

There are a small number of architects currently interested in the subject of crochet at is relates to architecture. The following publications, explorations and thesis projects can be used as a resource to see how crochet is being used within the capacity of architecture.

- Emergent Explorations: Analog and Digital Scripting - Alexander Worden
- Research and Design: The Architecture of variation - Lars Spuybroek
- YurtAlert - Kate Pokorny

Graffiti

In the past few years, a practice called yarn bombing, or the use of knitted or crocheted cloth to modify and beautify one's (usually outdoor) surroundings, emerged in the US and spread worldwide. Yarn bombers sometimes target existing pieces of graffiti for beautification.

Bobble (knitting)

In knitting, a bobble is a localized set of stitches forming a raised bump. The bumps are usually arranged in a regular geometrical pattern (e.g., a hexagonal grid) or may be figurative, e.g., represent apples on a knitted tree.

9

THE ART OF HAND KNITTING

Hand knitting is a form of knitting, in which the knitted fabric is produced by hand, using needles. While the origins of knitting are unclear, we know it has been practised in many different parts of the world, over many centuries, producing objects of great beauty as well as items fulfilling practical needs. When done by hand, it has used simple tools, such as hand-carved sticks of wood, bone, quill and ivory or metal wires and fine steel knitting needles that were commonplace in the 19th century.

Most filling knits (that includes plain, rib, purl pattern) can be made by hand or machine, although commercial fabrics are generally machine-made. Basic stitches are the knit stitch, a loop passed through the front of the preceding loop, and the purl stitch, drawn through the back.

With the fast growing popularity of knitting, among the younger generation there is a growing need for a site which details the latest styling knitting patterns. Knitting brain achieves this and provides some very helpful calculators which enable you to quickly do yarn substitution calculations.

KNITTING PATTERNS AND CALCULATORS

There is also a personality test which is quite fun. It's worth doing just to find out what sort of knitter you really are.

Knitting brain is on the fastest growing knitting resources and are sure to have the tip or answer to the problem you are facing. They also cover the knitters enemy - the wooly bear and how to identify this wool loving house dwelling pest.

Finger knitting, also known as finger weaving, is a great way to spend your free time when you need something fun to do. And when you're finished, you'll have a beautiful rope of yarn that you can use for lots of different things, such as keyring rope, hair decorations, a belt, or bag handles.

Knitting is something we are all familiar with, but surprisingly we know little about when and where the art of hand knitting started. This is because yarn fibers are biodegradable. One thing we do know is that the oldest form of knitting (crossed knitting) was practiced since before the birth of Christ. A pair of knitted socks discovered in Egyptian tombs from the 3rd to 6th centuries A.D. is the earliest archaeological evidence of knitted garments. The earliest knitting needle is a brass rod that dates back to the early Iron Age. Spinning wool started around 4000 B.C. near the Mediterranean Sea. The first wool factory in England was build by the Romans in 50 A.D. The kind of knitting that we are used to was not practiced till quite recently, however.

The oldest form of knitting is crossed knitting, also known as single-needle knitting and pseudo knitting. In crossed knitting, the stitches are rotated a half turn instead of aligning vertically. This knitting method was highly developed by the Nazca culture in Peru (100 B.C.-A.D. 700) in the fringes of their woven cloths. Frequent color changes in these fringes were used to create intricate human and animal figures.

The origins of knitting are hard to trace but there are several theories. Some people believe that knitting started in Persia, others that it started in Israel, Jordan and Syria. Still others believe that it started in the mountains of North Africa, or even Japan or China. Some people believe that knitting grew out of the knitting of fish-nets by men.

Some socks and other items made with the crossed-knitting technique have been found in Egyptian burials possibly dating as far back as the 4th or 5th century B.C. There have been knitted socks found in Egyptian tombs (3rd and 6th centuries A.D., knitted pieces found at the Dura-Europas site near the Euphrates River (about A.D. 200), and sandal socks apparently from Saudi Arabia (about A.D. 350). It is possible however that these earliest socks were worked in nalebinding; which is an ancient craft which often looks very similar to true knitting, and could be confused with true knitting by archaeologists with no training in the history of needlework. A pair of patterned cotton socks from Egypt, dating from A.D.1100 may be the earliest example of "true" knitting or that may be nalebinding too. Socks and stockings were the first items to be knitted because knitting was ideally suited for shaping a garment to fit the foot at a time when fabric and sewing woven cloth were less flexible.

The Complete Encyclopedia of Stitchery by Mildred Graves Ryan says that most historians agree that knitting was probably spread by (probably male) Arabian sailors and merchants who traveled throughout the Mediterranean. Many people believe that knitting was first invented by Arabian nomads who carried the craft into Egypt, probably in the 5th century A.D. Knitting was then carried through North Africa, and into Spain. From Spain, traveling Catholics picked it up and spread it quickly throughout Europe.

It was not until the early 14th century that we have the first references to true knitting in Europe. At that time, the purl stitch was unknown, which meant that in order to produce plain knitting, people had to knit in the round and then cut it open if they needed to. The first reference to purl stitch was not until the mid 16th century, but the knowledge of how to do it may have preceded that a little.

Although nobody knows exactly where true knitting started, it would seem that knitting was probably spread by Arabian sailors and merchants who traveled throughout the Mediterranean. Then, evidently, traveling Catholics quickly spread it throughout Europe. True knitting, as we know it, is a relatively recent craft. Crafts similar to true knitting, such as cross-knitting and nalebinding, however, have a very long history, dating back to before the birth of Christ. They evidently were practiced in a lot of different countries and cultures too, from Japan to Egypt to Peru. Knitting clearly filled a need in people's making of garments which is still continuing and growing.

Hand knitting is a special case of knitting, in which the knitted fabric is produced by hand. Weft-knit fabrics can be divided into two types: those that have selvages (side edges) and those that are tubes, where the side edges have been joined. The former are knit using "flat knitting", whereas the latter are knit using "circular knitting", also known as "knitting in the round".

In flat knitting, the hand-knitter generally knits from right-to-left on one side of the fabric, turns the work (over), and then knits right-to-left back to the starting position. Usually one side of the fabric is considered the right side, the one that faces outwards for viewing; the side that faces inwards, towards the body, is known as the wrong side. Thus, flat knitting involves knitting one row on the right side, then one row on the wrong side, etc.

Stitches knit on the wrong side are reversed in appearance; for example, a knit stitch carried out on the wrong side will appear as a purl stitch on the right side, and vice versa. Thus, the uniform stockinette fabric requires that the hand-knitter knit all the stitches on the right side, and purl all the stitches on the wrong side.

For comparison, garter-stitch fabric is produced if the knitter knits (or purls) every stitch in every row, regardless of which side is being worked.

In circular knitting, the knitter generally knits everything from one side, usually the right side. Circular knitting is usually carried out on a single circular needle, although this becomes more difficult as the circumference of the tube gradually shrinks. In such cases, the knitter can resort to a variety of alternative techniques, such as double-pointed needles, knitting on two circular needles, a Möbius strip-like "magic needle" approach (commonly known as "Magic Loop"), or careful use of slip-stitch knitting or equivalently double knitting to knit the back and front of the tube.

The Basics of Hand Knitting

Creating a slip knot– is the first step to begin knitting. A knot is the very first stitch a knitter creates in order to come up with a number of stitches. It fastens the yarn to your needles so you can cast on the rest of the stitches. Here's a step-by-step guide from WikiHow:

1. Pick up the yarn 6-8" from the end.
2. Twist your fingers clockwise to form a loop.
3. Expand the loop as you place your two fingers inside.
4. Pull the tail and tight knot halfway.
5. Insert your knitting needle inside the loop and pull both ends to create a tight knot.

Casting on – the second step in knitting. When you cast on, you create a series of stitches in order to create your chosen craft. There are three methods in casting:

Single Cast On– Also called the thumb cast on or backward loop cast on. This style is ideal for the beginners since it is the easiest way of casting stitches.

Long Tail Cast On– Also called the double cast. One of the most popular and quickest cast on methods. It creates a sturdy yet flexible set of stitches for most knitted projects. The only challenge for this type is the proper measurement. You have to make sure that you leave enough of the tail to cast all your stitches.

Knitted Cast On– Also known as two-needle cast on. This method of casting creates two stitches out of one. It's a great technique to try since you'll eventually learn the other steps in knitting once you get used to this style. Through knitted cast on, you can add extra stitches in the middle of a row. If you had problems in measuring the long tail cast on, this method can serve as your alternative as well.

Knit Stitch- Also known as the garter stitch. It is the process of pulling up a yarn through an existing loop on a needle and pulling it through on the back.

Purl Stitch- It is the method of inserting the needle from the top of a loop and pulling a new loop up and through the first stitch. It's almost the same with knit a stitch, yet they differ with some of the motions. A purl stitch simply looks like the back of a knit stitch. Each row that a knitter purls will create a bumpy

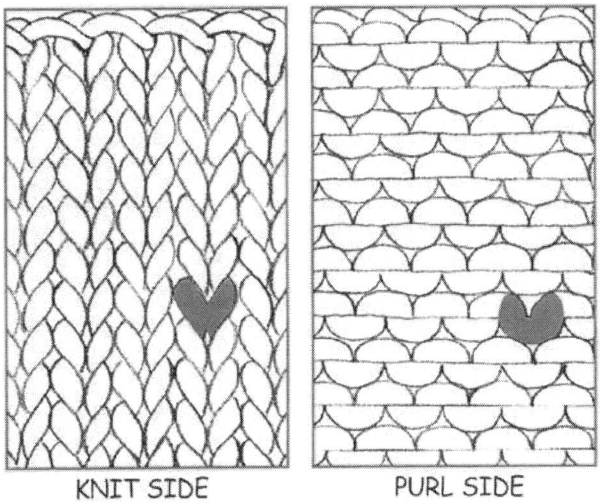

Stockinette Stitch - Simply alternating a knit row with a purl row. Usually applicable in knitting scarves, socks, sweaters, blankets, and hats. (To learn the step-by-step process in knitting a stockinette stitch, click here.)

Ribbing Stitch– Purling back and forth within the same row creates this kind of technique. Because of its flexibility and ability to bounce back, ribbing is widely used to make up hems, necks, and cuffs of sweaters. It is also used to create a slim, body-hugging effect in most garments.

Seed Stitch– A sequence of alternating knit and purl stitches done horizontally and vertically. The purl bumps look like scattered seeds in the fabric, making it more creative in texture and in pattern.

Cable Stitch– Creating a cable row is also known as the turning row. This method is usually applied on a stockinette stitch background. It is not necessarily made on every row. For instance, a 4-stitch cable will have plain rows of knitting before a cable row, and so on.

Casting Off – also known as Binding Off, this method is the last step in knitting. Its main purpose is to make an edge so the stitches that you've knitted won't unravel. (Check out the step-by-step process on casting off, here.)

Fixing Common Mistakes

As a beginner, of course, there's a possibility that you'll commit some mistakes in creating your first few stitches. But don't worry. We're here to help you out.

Dropped Stitches – A very common mistake in knitting. It happens when a stitch just slips of your needle without you even recognizing it. This causes a run in the knitted material like a run in a pair of stockings. Fortunately, we've got a solution for this! Click here.

Twisted Stitches– Twisting stitches can either be done by accident or for a creative purpose. To know how to fix a twisted stitch, click here.

We've got you some knitting patterns from Simpleknitting.com that you'll surely enjoy working on. So grab your needles and yarn now. Let's try this!

Knitted Hat– using the Garter stitch method

Neck Warmer– using the Knit stitch method

Scarf– using the Seeded Rib stitch method

Ruffle Wrist Warmers– using the Stockinette stitch method

Shawl – using the Knit stitch and Purl stitch method

Types of Hand Knitting

Flat knitting

Flat knitting uses two straight needles to make generally two-dimensional (flat) pieces. Flat knitting is usually used to knit flat pieces like scarves, blankets, afghans, and the backs, fronts and arms of sweaters (pullovers).

In flat knitting, generally stockinette stitch, the hand-knitter knits from right-to-left on one side of the fabric, turns the work (over), and then purls right-to-left back to the starting position. Usually, the smooth side of the fabric is considered the right side, the one facing outwards for viewing; and the side that faces inwards, towards the body, the ridged side, is known as the wrong side. Thus, flat knitting involves knitting each row on the right side, then purling each row on the wrong side, etc. If each row is knit (no purls) this creates garter stitch, which has the same appearance on both sides and creates horizontal ridges offset by valleys, rather than a knit and purl side. Patterned stitching, such as cables, can be accomplished with either flat knitting, or in the round, however, the technique must follow the desired pattern.

Circular knitting

Circular knitting (also called "knitting in the round") creates a seamless tube. Knitting is worked in rounds (the equivalent of rows in flat knitting).

Originally, circular knitting was done using a set of four or five double-pointed knitting needles. Circular needles were later invented making this type of knitting easier. A circular needle resembles two short knitting needles connected by a cable of varying length between them. A circular knitting needle with a long cable can be used in place of straight needles to create larger flat-knitted pieces of fabric. Both types of circular knitting are used in creating pieces that are circular or tube-shaped, such as hats, socks, mittens, sleeves, and entire sweaters.

In circular knitting, the hand-knitter generally knits everything from one side, usually the right side. Circular knitting is usually carried out on a single circular needle. In such cases, the knitter can resort to a variety of alternative techniques, such as double-pointed needles, knitting on two circular needles, a Möbius strip-like "magic needle" approach (commonly known as "Magic Loop"), or careful use of slip-stitch knitting or equivalently double knitting to knit the back and front of the tube.

Fabric Finishing:

- Felting: Felting is the hand-knitters' term for fulling, a technique for joining knitted or woven animal-yarn fibres. The finished product is put in hot water and agitated until it starts to shrink. The end result typically has a felt-like appearance but has reduced dimensions. Bags, mittens, vests, socks, slippers and hats are just a few items that can be felted.
- Needle felting: Needle felting is a technique used to add decoration to a knitted or felted piece, where raw roving is applied using a very sharp barbed felting needle by repeatedly piercing the roving and background together. Once washed in hot water, the appliqued decoration is fused with

the background. Felted knitting can be cut with scissors without concern about fraying.

Process

There are many hundreds of different knitting stitches used by knitters. A piece of knitting begins with the process of casting on, which involves the initial creation of the stitches on the needle. Different methods of casting on are used for different effects: one may be stretchy enough for lace, while another provides a decorative edging. Provisional cast-ons are used when the knitting will continue in both directions from the cast-on. There are various methods employed to cast on, such as the "thumb method" (also known as "slingshot" or "long-tail" cast-ons), where the stitches are created by a series of loops that will, when knitted, give a very loose edge ideal for "picking up stitches" and knitting a border; the "double needle method" (also known as "knit-on" or "cable cast-on"), whereby each loop placed on the needle is then "knitted on," which produces a firmer edge ideal on its own as a border; and many more. The number of active stitches remains the same as when cast on unless stitches are added (an increase) or removed (a decrease).

Most Western-style knitters follow either the English style (in which the yarn is held in the right hand) or the Continental style (in which the yarn is held in the left hand). These styles show various regional differences. English knitting can be split up into, among others, throwing, flicking, and lever knitting. Continental knitting also shows a similar range of styles. Stitches can be produced by wrapping the needle around the yarn) or by simply picking it through (often called "Continental picking").

There are also different ways to insert the needle into the stitch. Knitting through the front of a stitch is called Western knitting. Going through the back of a stitch is called Eastern knitting. A third method, called combination knitting, goes through the front of a knit stitch and the back of a purl stitch. In Scandinavia, but especially in Norway, the purl stitch is produced with yarn held in back and held as close to the needles as possible. This shifts the tensioning of the yarn away from the fingers and on to the needle.

Once the knitted piece is finished, the remaining live stitches are "cast off". Casting (or "binding") off loops the stitches across each other so they can be removed from the needle without unravelling the item. Although the mechanics are different from casting on, there is a similar variety of methods.

In knitting certain articles of clothing, especially larger ones like sweaters, the final knitted garment will be made of several knitted pieces, with individual sections of the garment knit separately and then sewn together. Seamless knitting, where a whole garment is knit as a single piece, is also possible. Elizabeth Zimmermann is probably the best-known proponent of seamless or circular knitting techniques. Smaller items, such as socks and hats, are usually knit in one piece on double-pointed needles or small diameter circular needles. (See Circular knitting.)

HOW TO FINGER KNIT?

Here's how to do it. Loop the yarn around your thumb of your non-dominant hand. Don't tie a knot. The tail should be on the back of your hand, and you should press the side of your thumb against your index finger to hold it tightly.

Weave the yarn over your fingers four times. Take the other, "working" end of the yarn and pull it over your index finger, keeping the tail tightly gripped between the side of your thumb and index finger. Go under your middle finger, over your ring finger, and under your pinky. Loop around, going over the pinky, under the ring finger, over the middle finger, and under the index finger. Do this sequence one more time so that each finger has two complete loops on it.

Pull up the bottom loops. Take the bottom "line" of yarn on your pinky and pull it over your finger, taking it up above the first line. That loop of yarn should now be on the back of your hand. Repeat with all your fingers, moving from pinky to middle finger.

- When you reach your index finger, simply move the tail that was between your thumb and index finger between your index and middle fingers. It should be hanging down the back of your hand. This counts as the first "bottom loop" for your index finger.

Weave the yarn in and out of your fingers again until there is another line. Pull the bottom loops up again. Continue for as long as you want or until you run out of yarn. A nice rope-like structure of yarn should form at the back of your hand.

Finish off. When your rope is as long as you want it, stop when you have one loop on each finger.

- Take the loop from your pinky, and put it on your ring finger. Pull the bottom loop up from your ring finger and over to the back of your hand.
- Move the loop from your ring finger to your middle finger. Again, pull up the bottom loop over to the back of your hand.
- Move the loop from your middle finger to your index finger. Pull up the bottom loop again. You should be left with only one loop on your index finger now.

- Slip the loop off your index finger.

Cut your working yarn a few inches away from the loop, and pull the tail through the loop. Tug a few times to tighten. You can also pull on the original tail to tighten it as well.

Tie the ends of the rope in a square knot to secure.

- You can tighten the rope every few stitches by pulling on the end of the rope.
- It is much better to do this all at the one time, or you might lose the rhythm and forget where you were.
- You can place the loops on your fingers onto a pencil or straw to hold your place so you can come back to your weaving another time.
- You can make scarves, belts, bracelets, headbands, pot holders, and more.
- If you want to make a thinner, faster strand, simply do the above method using only one finger instead of four.
- If you see loose loops in your rope, don't cut it. If the loop is close to your knuckles, just pull out the stitches back to the problem. Then redo that part.
- If you suffer from Repetitive Stress Injury, remember to take frequent breaks.

STITCHES

There are well-nigh an infinite number of possible combinations of knitting stitches, the favorites of which have been collected into stitch treasuries. A piece of knitting begins with the process of casting on (also known as "binding on"), which involves the initial creation of the stitches on the needle. Different methods of casting on are used for different effects: one may be stretchy enough for lace, while another provides a decorative edging - Provisional cast-ons are used when the knitting will continue in both directions from the cast-on.

The Basic Stitches

The knit stitch (k)

1. With the yarn at the back of the work, insert the right hand needle from left to right through the front of the first stitch on the left hand needle.

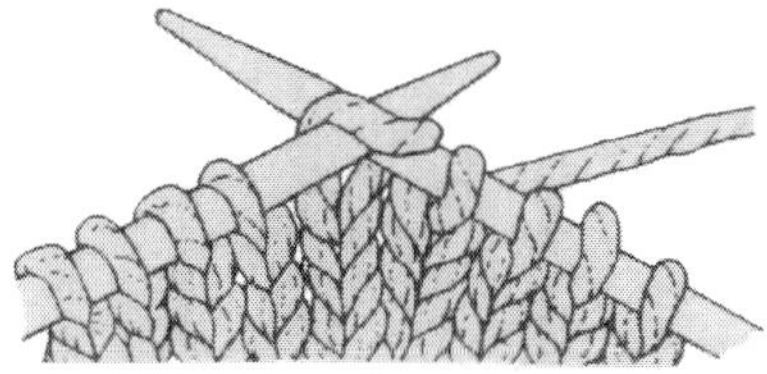

2. Wind the yarn over the right hand needle.

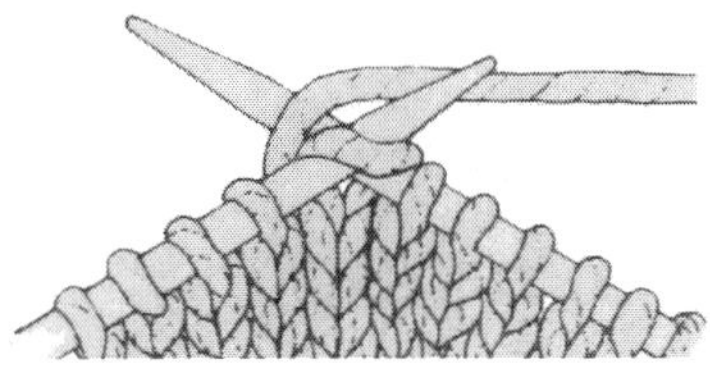

3. Pull through a loop.

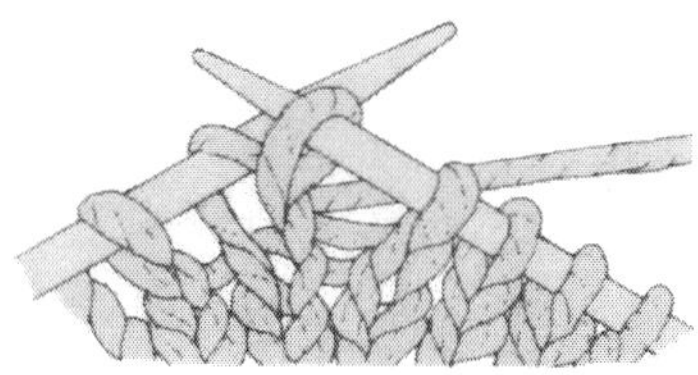

4. Slip the original stitch off the left hand needle.

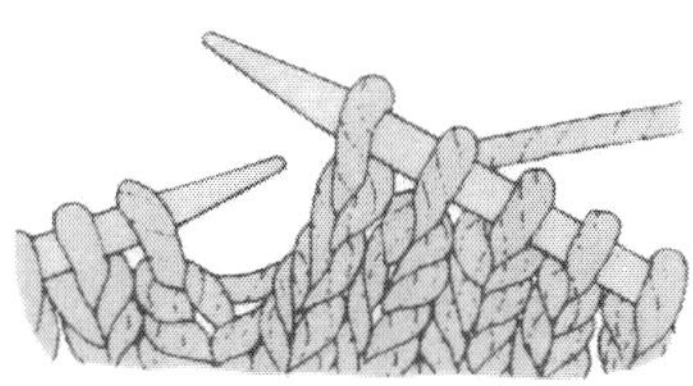

Repeat until all the stitches have been transferred from the left to the right hand needle.

To knit into the back of a stitch (KB1)

This technique is sometimes used to create a twisted rib effect, as well as some textured stitch patterns. Insert the needle into the back of the stitch on the left hand needle.

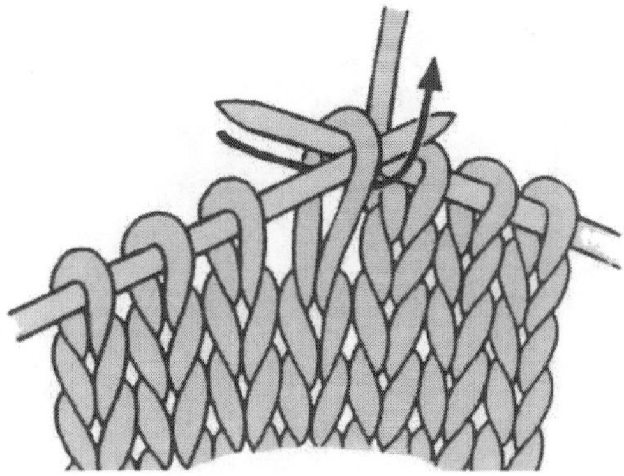

then follow steps 2 to 4 for the knit stitch.

The purl stitch (p)

1. With the yarn at the front of the work, insert the right hand needle from right to left through the front of the first stitch on the left hand needle.

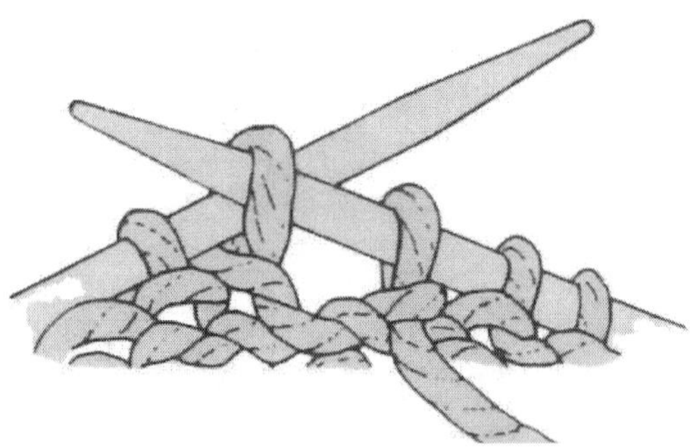

2. Wind the yarn round the right hand needle.

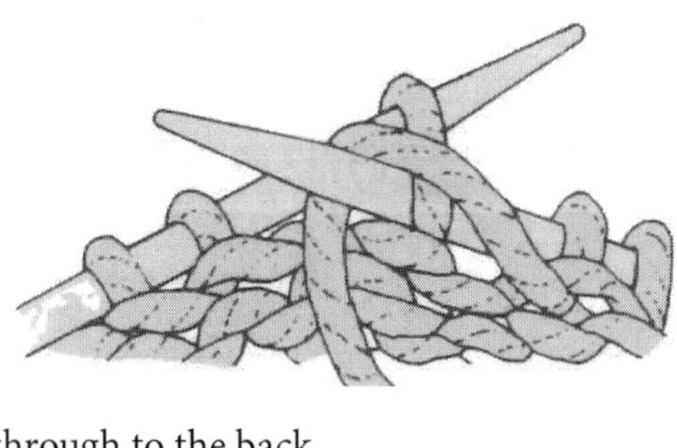

3. Draw a loop through to the back.

4. Slip the original stitch off the left hand needle.

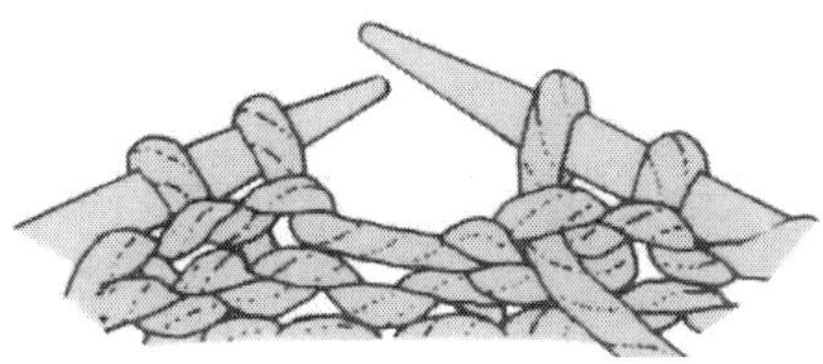

To purl into the back of a stitch (PB1)

Insert the needle into the back of the stitch on the left hand needle from behind and bring the point of the right hand needle through to the front.

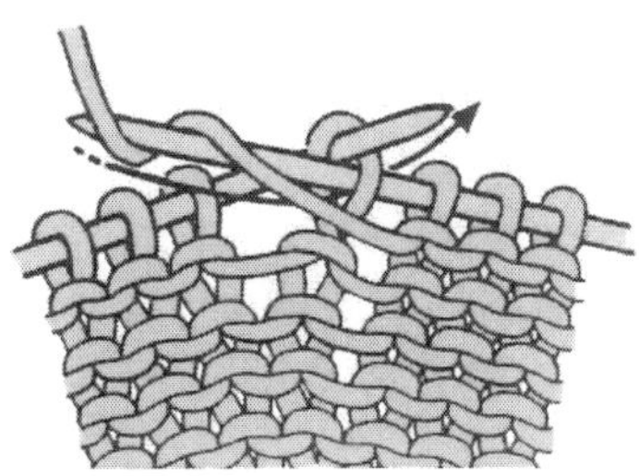

then follow steps 2 to 4 for the purl stitch.

The garter stitch

The garter stitch is formed by working every row as a knit row, the first row usually being the right side

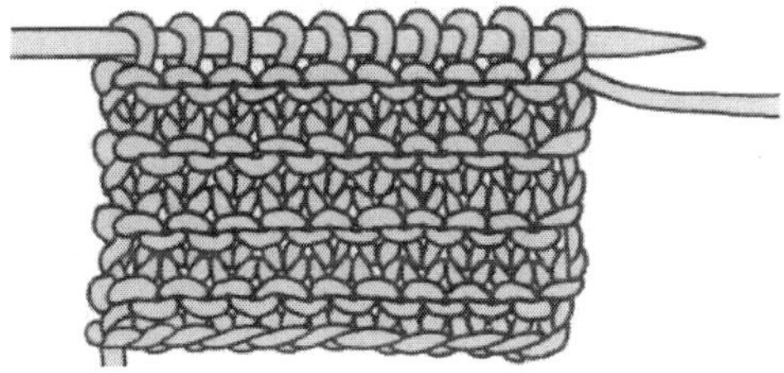

The same effect can also be achieved by working every row as purl.

Stocking stitch

Stocking stitch is made up of one row of knit stitches followed by one row of purl stitches, starting with a knit row, which is the right side of the work.

Reverse stitch

Reverse stocking stitch is made up of one row of purl stitches followed by one row of knit stitches, starting with a purl row, which is the right side of the work.

Ribbing

Ribbing is an elastic fabric which is often used for garment edgings. The two most usual types are 1x1 rib, which is formed by alternating 1 knit stitch and 1 purl stitch and 2x2 rib which is formed by alternating 2 knit stitches and 2 purl stitches. Care must be taken to purl the stitches which were knitted on the previous row and vice versa

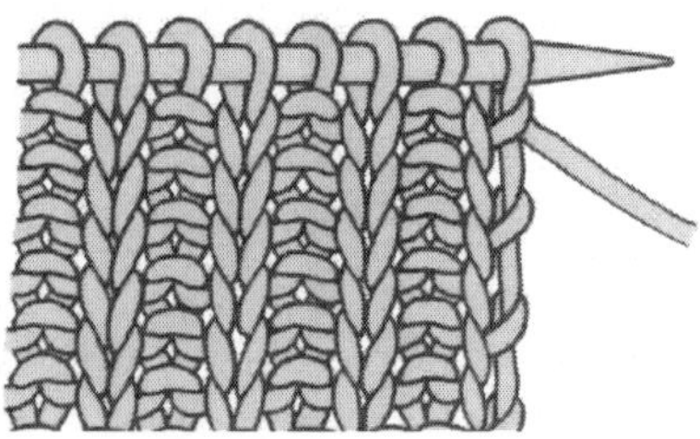

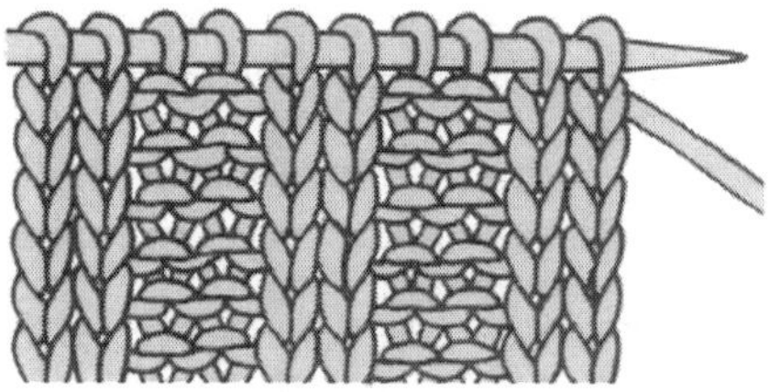

Knit one Below (K1B)

This stitch is used in Fishermen's rib. Insert the right hand needle into the next stitch but in the row below the stitch on the left hand needle. Then knit the stitch as normal.

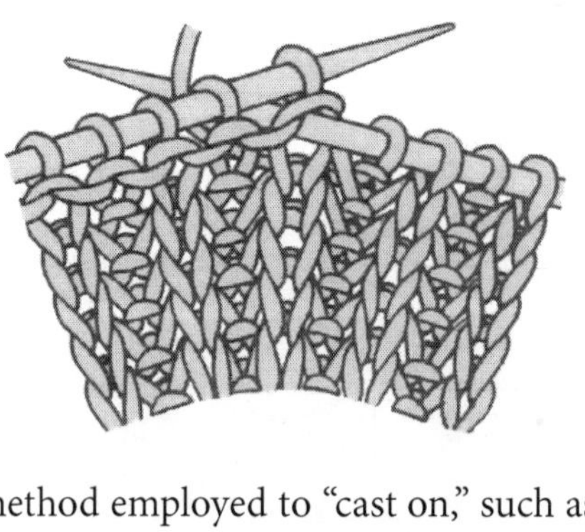

There are various method employed to "cast on," such as the "thumb method" (also known as "slingshot" or "long-tail" cast-ons), where the stitches are created by a series of loops that will, when knitted, give a very loose edge ideal for "picking up stitches" and knitting a border; the "double needle method" (also known as "knit-on" or "cable cast-on"), whereby each loop placed on the needle is then "knitted on," which produces a firmer edge ideal on its own as a border; and many more. The number of active stitches remains the same as when cast on unless stitches are added (an increase) or removed (a decrease).

Most Western-style knitters follow either the English style (in which the yarn is held in the right hand) or the Continental style (in which the yarn is held in the left hand). A third but less common method, called combination knitting, may also be used.

Once the knitted piece is finished, the remaining live stitches are "cast off." Casting (or "binding") off loops the stitches across each other so they can be removed from the needle without unravelling the item. Although the mechanics are different from casting on, there is a similar variety of methods.

In knitting certain articles of clothing, especially larger ones like sweaters, the final knitted garment will be made of several knitted pieces, with individual sections of the garment knit separately and then sewn together. Seamless knitting, where a whole garment is knit as a single piece, is also possible. Elizabeth Zimmermann is probably the best-known proponent of seamless or circular knitting techniques. Smaller items, such as socks and hats, are usually knit in one piece on double-pointed needles or circular needles. (SeeCircular knitting.)

SOCIAL ASPECTS

One of the earliest known examples of knitting was finely decorated cotton socks found in Egyptin the end of the first millennium AD. The first knitting trade guild was started in Paris in 1527. With the invention of the knitting machine, however, knitting "by hand" became a useful but non-essential craft. Similar to quilting, spinning, and needlepoint, knitting became a social activity.

Hand-knitting has gone into and out of fashion many times in the last two centuries, and at the turn of the 21st century it is enjoying a revival. According to the industry group Craft Yarn Council of America, the number of women knitters in the United States age 25-35 increased 150% in the two years between 2002 and 2004. While some may say knitting has never really gone away, this latest reincarnation is less about the make do and mend of the 1940s and 50's and more about making a statement about individuality as well as developing an innate sense of community. Additionally, many contemporary knitters have an interest in blogging about their knitting, patterns, and techniques.

There are now numerous groups that are not only growing individually, but also forming international communities. Communities also exist online, with blogs being very popular, alongside online groups and social networking through mediums such as Yahoo! Groups, where people can share tips and techniques, run competitions, and share their patterns. More people are finding knitting a recreation and enjoying the hobby with their family. Knitting parties also are becoming popular in small and large communities around the U.S. and Canada.

Psychological and meditative aspects

The oral histories of many knitters have been collected, and suggest that hand-knitting is often associated with compassion. "I knit love into every stitch" is a common refrain.

Knitting especially large or fine garments such as sweaters can require months of work and, as gifts, may have a strong emotional aspect. The so-called sweater curse expresses the experience that a significant other will break up with the knitter immediately after receiving a costly hand-knit gift such as a sweater.

A significant minority of knitters claim to have experienced the sweater curse; a recent poll indicated that 15% of active knitters say they have experienced the sweater curse firsthand, and 41% consider it a possibility that should be taken seriously. Although sometimes labeled a "superstition", the sweater curse is not treated in knitting literature as anything paranormal.

Hand-knitting is generally relaxing and repetitive, Some practitioners have noted that these factors, combined with its compassionate nature, make hand-knitting well-suited for meditational or spiritual practice.

Knitting is sometimes featured in literature. Knitting and its techniques may be used as a metaphor; its meditative and spiritual aspects may be emphasized; it may signal various types of domesticity; or it may be used for dramatic irony, as when an apparently harmless knitter proves deadly and implacable.

Examples from 19th century novels include Madame Thérèse Defarge in Charles Dickens' A Tale of Two Cities, Anna Makarovna in Leo Tolstoy's War and Peace, various characters in Jane Austen's novels and Miss Ophelia in Harriet Beecher Stowe's Uncle Tom's Cabin. Several characters in Virginia Wolff's novels are knitters. In the first decade of the 21st century, knitting has been a key element in several novels and even murder mysteries.

KNITTING FRAMES (BOARDS,

Rings and Spools)

It is also possible to knit using rigid frames that hold the active stitches. A new row is added by wrapping the yarn about the posts of the frame; the previous loops are pulled over the new ones, to suspend the former from the latter. The shape and number of posts of the frame can determine the type of fabric produced. Knitting can be done in reverse using metal wire by using plaited stitches.

Ancillary tools

Various tools have been developed to make hand-knitting easier. Tools for measuring needle diameter and yarn properties have been discussed above, as well as the yarn swift, ballwinder and "yarntainers". Crochet hooks and a darning needle are often useful in binding off or in joining two knitted pieces edge-to-edge. The darning needle is used in duplicate stitch (also known as Swiss darning), while the crochet hook is also essential for repairing dropped stitches and some specialty stitches such as tufting.

Other tools are used to prepare specific ornaments include the pompom tree for making pompoms conveniently. For large or complex patterns, it is sometimes difficult to keep track of which stitch should be knit in a particular way; therefore, several tools have been developed to identify the number of a particular row or stitch, including circular stitch markers, hanging markers, extra yarn and counters.

A second potential difficulty is that the knitted piece will slide off the tapered end of the needles when unattended; this is prevented by "point protectors" that cap the tapered ends. Another problem is that too much knitting may lead to hand and wrist troubles; for this, special stress-relieving gloves are available. Finally, there are sundry bags and containers for holding knitting projects, yarns and needles.

Pocket Scarf

If you work somewhere that's a little cold, a pocket scarf is a great winter accessory. Like a regular scarf, it will keep you warm as you go about your day, but when you're up and about, chatting with coworkers or at meetings, you can use the pockets to keep your hands warm.

The little pockets are also a great place to stash a travel size tissue case, your lip balm or anything else you need to keep close at hand in the winter months.

Materials

- Three skeins of Knit Picks Twirl (I used color Fog), or about 200 yards of a similar bulky weight nubby yarn
- One pair size 11 US needles
- Scissors
- Tape measure or ruler

- Yarn needle

Gauge

10 stitches and 17 rows per 4 inches (2.5 stitches and 4.25 rows per inch) in garter stitch. Gauge is not critical.

Size

Finished size is 6 inches wide and 60 inches long with 10-inch pockets.

Instructions:

1. Cast on 15 stitches.
2. Knit in Garter Stitch for 80 inches.
3. Bind off.
4. Cut yarn and weave in ends, or use them to help sew pockets.
5. Fold the last 10 inches of each end of the scarf over on itself and sew down using working yarn, or a smoother yarn if that's easier for you.

Make it Your Own: This project is designed so that the scarf can be worn straight around the neck and the pockets are in the right place to be useful when standing up.

To make this scarf fit you, put a tape measure around your neck as you would wear a scarf, measure where the tips of your fingers fall and add 10 inches to each side.

The yarn used in the pictured project is an alpaca, wool, nylon blend, which makes it very warm and a little shiny, but you can use any highly textured yarn you like.

KNITTING PATTERNS FOR THE WARMEST SCARVES

People knit different garments for different reasons, but when the coldest days of winter set in, you want a scarf knitting pattern that you know is going to make a nice, warm scarf for you. These are the best knitting patterns for the warmest scarves, projects that will keep you cozy no matter what the weather throws at you.

This is probably the warmest scarf knitting pattern available here at About.com Knitting and with very good reason: the scarf is a double thickness so that the back side of the cable doesn't show. This makes for slower knitting than a one-sided scarf of the same length would be, but the warmth factor is well worth it.

Using lots of different strands of yarn held together makes for a big, quick knitting project, and when those yarns are all wool, you also get a very warm scarf. That's the thinking behind the gorgeous and super-easy Favorite Color Scarf. Pick up a few skeins of your favorite colors, some size 50 needles and in no time you'll have a huge scarf to help stave off the cold.

Another way to get a warm scarf knit quickly is to use a really bulky yarn, as in the Basketweave Scarf. This one can literally be whipped up in a few hours, and the super bulky yarn is warm and comfortable. The basketweave pattern adds interest in the knitting and the wearing.

Your scarf can keep you extra warm when it also serves as a place to put your hands to keep them extra warm. The pockets in this Pocket Scarf are simply made by folding over the ends of the scarf and sewing the side seams, and they're a great place to stash a little packet of tissues or some lip balm if you work in a cold office and wear your scarf all day long.

This Women's Sampler Scarf is among the warmest scarves because it's so darn wide. You can fold the scarf in half and have a double thickness of warmth around your neck (or cast on fewer stitches and make the scarf longer, but it's really the part around your neck that helps keep the wind out the best).

There's nothing inherently warm about the Horizontal Scarf knitting pattern, other than the length of the scarf, which allows you to wrap it around your neck multiple times. But knitting a scarf horizontally means you can make it as wide as you want, too; just stop when you think you've got enough width to keep yourself cozy.

A cowl is not, strictly speaking, a scarf, but it is something that wraps around your neck and can help keep you warm. The Candle Flame Cowl is a beauty, worked in wool yarn that's sure to keep you warm and with a design that's sure to turn heads no matter what color you knit it in.

It may be hard to believe that such a tiny project can be warming, but the Razor Shell Scarflette is a great choice for a base layer (a scarf you tuck into your shirt, maybe?). That's because it's made out of cashmere, which is super-soft and, yes, really warm. You might still want a bulkier scarf on really cold days, but this bit of luxury against your skin will almost make the cold days worthwhile.

Basketweave Scarf

Winter Warmer: When it's really cold outside, you don't want to mess around with a short or skimpy scarf. You need a scarf that will really keep you warm, a big, warm, wooly scarf.

It doesn't hurt when that scarf also looks good and is quick and easy to knit. That's what the basketweave scarf is all about.

Basketweave is an easy stitch pattern with many variations that combines knits and purls into a fabric that looks woven. This basketweave pattern is based on a simple eight-row repeat.

Materials

- About 200 yards of super-bulky wool or wool-blend (I used two skeins of Lion Brand Wool Ease Thick and Quick in Fisherman)
- Size 13 needles
- Scissors
- Crochet hook or tapestry needle

Size

Finished scarf is 60 inches long and almost eight inches wide. Gauge is not critical.

Instructions:

1. Cast on 19 stitches using your favorite method.
2. Knit in the following pattern:

- Row one: Knit across.
- Row two: Knit four, purl three, knit five, purl three, knit four.
- Row three: Purl four, knit three, purl five, knit three, purl four.
- Row four: Repeat row two.
- Row five: Knit across.
- Row six: Purl three, knit five, purl three, knit five, purl three.
- Row seven: Knit three, purl five, knit three, purl five, knit three.
- Row eight: Repeat row six.

Repeat pattern to desired length or 60 inches.

Bind off.

Cut excess yarn if needed and weave in ends.

Fun With Basketweave: Basketweave is a lovely stitch pattern, but it is not reversible. That means this scarf has an obvious front and back that aren't the same. The back isn't ugly by any means--it has an interesting pattern of raised squares that would make a nice scarf in its own right. If you like the look of a scarf that looks the same front and back, don't make this pattern. Otherwise, it is a fun pattern that's easy to work and very satisfying in how quickly it knits up. Just one pattern repeat will give you about two and a half inches of scarf and takes 10 minutes or less to knit.

Beginner's Garter Stitch Scarf

The Perfect Pattern for Beginners: If you're new to knitting and are contemplating your first project, I would like to recommend this one, the beginner's garter stitch scarf. This is an easy pattern because all you need to know is how tocast on, make the knit stitch and bind off. It literally cannot get any easier than this.

More experienced knitters should take a look at this pattern, too. It makes a handsome scarf that you can complete in just a couple of hours--perfect for last-minute gift giving.

Materials

- One skein super-bulky yarn, around 100 yards (I used Lion Brand Wool Ease Thick & Quick in Denim)
- Size 13 needles, straight
- Crochet hook

This makes a scarf about five inches wide and 60 inches long, depending on the yardage of your chosen yarn.

Instructions:

1. Cast on 12 stitches.
2. Knit every row until you have about a yard of yarn left.
3. Bind off.
4. Use your crochet hook to weave in the ends at the top and bottom.

See, I told you it was easy!

Make the Project Your Own: This is a wonderful all-purpose scarf pattern. You can make it with any kind of yarn or with a couple of different kinds of yarn held together.

If you want to use a different weight of yarn, check on the label to see what sized needles are recommended and use them instead of the size 13s. You might also need a few more stitches if you're using lighter weight yarn in order to get a nice width. Or you can make a super-skinny scarf. That's fun, too.

Warmth in No Time

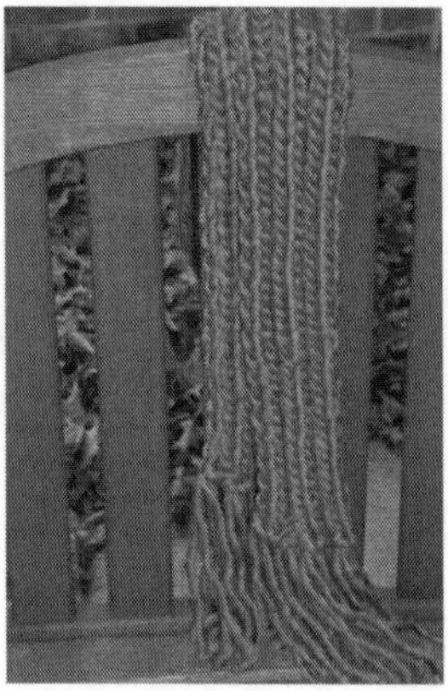

Bulky Ribbed Scarf.

(c) Sarah E. White, licensed to About.com, Inc.

This bulky ribbed scarf is a great way to use some of the bulky yarn in your stash that you might not have known what to do with. A scarf is a great way to use bulky yarn because when it gets too warm you can just take it off.

This scarf, worked in an alpaca-wool blend, is big and long enough to keep you warm on even the coldest days. You can add the fringe if you like or leave it off and knit the scarf longer or not as you like.

Materials

- 3 skeins Blue Sky Alpacas Bulky (I used color 1219, Atlantis), or 135 yards of the bulky yarn of your choice
- One pair size 15 US (10 mm) knitting needles
- Gauge:

Gauge is not critical, but I got 12 stitches and 9 rows per 4 inches in ribbing (3 stitches and 2.25 rows per inch).

Size

Finished scarf is 4 inches wide and 60 inches long without fringe (about 74 inches with fringe). If you omit fringe and knit until you're almost out of yarn, your scarf should measure about 75 inches.

Knitting the Scarf

1. Cast on 12 stitches.
2. Work in knit 2, purl 2 ribbing until piece measures about 60 inches, or to

length of your choice.

3. Bind off in pattern.
4. With remaining yarn, if desired, cut 24 pieces of yarn about 18 inches in length. For each fringe, fold two pieces of yarn in half, use a crochet hook to pull them through the edge of the knitting and pull the ends through the loop. Do this 6 times on each end of the scarf (one for each section of the ribbing).
5. Weave in any remaining yarn ends.

Condo Knit Scarf

Quick and Breezy: Condo knitting is a fun and easy knitting technique that involves knitting alternating rows of a patten with widely different sized knitting needles.

This gives the finished fabric an open appearance without any fancy stitchwork. Using two needles that aren't incredibly far apart, as in this project, gives a look sort of like dropped stitches without any effort at all.

Materials

- One skein Deborah Norville Collection Serenity Sport Weight[/link from Premier Yarns (I used color 13, pink carnation), or about 230 yards of your favorite fine yarn
- One pair size 5 US knitting needles and one size 15 knitting needle
- Scissors and yarn needle

Gauge

17 stitches per 4 inches (4.25 stitches per inch) on smaller needles and 21 rows per 4 inches (5.25 rows per inch) in condo knitting (Garter Stitch). Gauge is not critical.

Size

Finished size is 9 inches wide and 48 inches long.

Instructions:

1. Using size 5 needle, cast on 30 stitches.
2. Knit one row.
3. Change to size 15 needle and knit 1 row.
4. Knit one row with size 5 needle.

5. Continue in this manner, changing needle size each row, until piece measures around 48 inches, or length of your choice, ending with a row on the size 5 needle.
6. Using the size 5 needles, bind off.
7. Cut yarn and weave in ends.

Make It Your Own:

I like the look of this scarf very wide, but you can cast on fewer stitches if you'd like something a little less substantial. And if you'd like bigger holes in your scarf, use a size 35 needle instead of the size 15 (that's more traditional) to make a much airier piece.

To wear, wrap around your neck as you would a scarf, or try draping it over your shoulders and closing with a pin for a shawl effect.

COLOR KNITTING TECHNIQUES AND METHODS

There are a wide variety of color knitting techniques that allow you to make basic knitting projects a lot more interesting. From knitting in stripes or using a multicolored yarn to complex entrelac and intarsia knitting projects, there are lots of ways to brighten up your knitting that aren't necessarily difficult. Here's a look at some of them.

Multicolored Yarn

Working with multicolored yarn is the easiest way to bring color to your projects because you don't even have to think about it; just keep knitting as normal.

Whether the yarn is variegated (a series of colors that are often random) or self-striping, knitting a project with this kind of yarn makes it look much more complicated than it is and adds a bit of pizazz even to plain stockinette stitch.

Knitting Stripes

Adding stripes to a knitting pattern is not that difficult, just remember to change colors at the end of a row or round. If you're knitting in stockinette stitch or a similar stitch and don't want the broken line where the colors change to show, start the new color on a knit row. Sometimes, however, you might want to use that line as a design element. Projects that include stripes can be a garter stitch striped baby blanket or a classic striped stocking hat.

Duplicate Stitch

Using duplicate stitch is an easy and fun way to add color to a knitting project after the fact. You can use as much or as little duplicate stitching as you want on a project, but be mindful that the stitching makes the work bulkier and a little stiffer than it would be without that extra layer of stitches.

Duplicate stitch is also a great way to add more personality to projects, as with the animal hat patterns.

Slip-Stitch Knitting

Slip-stitch knitting looks more complicated than it is because it uses multiple colors of yarn to make designs, but you are only working with one color at a time. Other stitches are slipped so that the color from the previous row shows on those stitches, making interesting patterns.

This technique is also known as mosaic knitting, which is actually a specific technique that was popularized by Barbara G. Walker. It involves designs that can be worked on any number of stitches and are charted, with the wrong side row worked in the same manner as the right side.

Fair Isle or Stranded Knitting

Stranded knitting, also sometimes referred to as Fair Isle, is a relatively easy way to work two colors into the same row of knitting.

Holding one yarn in each hand or dropping and picking up colors as you need them, the unused color is stranded across the back of the work, giving you a double-thick layer of knitting that's super warm.

This technique is good for simple patterns that are just a couple of stitches in width and repeat across a row.

Intarsia Knitting

Knitting intarsia, or picture knitting, is the way to add more complex, larger designs that don't cover the whole width of a knitting project. Each color is worked as a block, with a different strand of yarn used for each time a color appears in the project.

For example, if there's a red motif in the middle of two sections of brown knitting, you'll need three strands of yarn, one red and two brown, to work the pattern.

Entrelac Knitting

Entrelac does not necessarily have to be worked in color, but the woven patterns are traditionally knit using at least two colors.

One really fun way to work entrelac that's also quite easy is to use a multicolored or self-striping yarn to work the squares and triangles, which adds a lot of color with no extra work.

10

LACE KNITTING

Lace knitting is a style of knitting characterized by stable "holes" in the fabric arranged with consideration of aesthetic value. Lace is sometimes considered the pinnacle of knitting, because of its complexity and because woven fabrics cannot easily be made to have holes. Famous examples include the Orenburg shawl and the wedding ring shawl of Shetland knitting, a shawl so fine that it could be drawn through a wedding ring. Shetland knitted lace became extremely popular in Victorian England when Queen Victoria became a Shetland lace enthusiast. Her enthusiasm resulted i.a. in her choosing knitted lacework for presents; e.g. when in ca. 1897 the Queen gave a lace shawl as a present to American abolitionist Harriet Tubman. From there, knitting patterns for the shawls were printed in English women's magazines where they were copied in Iceland with single ply wool.

Some consider that "true" knitted lace has pattern stitches on both the right and wrong sides, and that knitting with pattern stitches on only one side of the fabric, so that holes are separated by at least two threads, is technically not lace, but "lacy knitting", although this has no historical basis.

Eyelet patterns are those in which the holes make up only a small fraction of the fabric and are isolated into clusters (e.g., little rosettes of one hole surrounded by others in a hexagon). At the other extreme, some knitted lace is almost all holes, e.g., faggoting.

Knitted lace with no bound-off edges is extremely elastic, deforming easily to fit whatever it is draped on. As a consequence, knitted lace garments must be blocked or "dressed" before use, and tend to stretch over time.

Lace can be used for any kind of garment, but is commonly associated with scarves and shawls, or with household items such as curtains, table runners or trim for curtains and towels. Lace items from different regional knitting traditions are often distinguished by their patterns, shape and method, such as Faroese lace shawls which are knit bottom up with center back gusset shaping unlike a more common neck down, triangular shawl.

Know how to decreases and increases. The most used ones are: knit two together (k2tog), make one (m1), slip slip knit (ssk), and yarn over (yo) which creates the holes in lace. You must learn this before you start practicing lace knitting, and also some other increases and decreases.

Choose your yarn. Lace knitting is made with thin, skinny yarns. Any color will do.

Choose your needles. Pick thin needles like size 2 knitting needles.

Search online for lace patterns. You can search knitting pattern central. com, or you can search for some lacepatterns on knitting magazines.

Begin your pattern with four rows of garter stitch, as most patterns call for. Even if you are just practicing, you still need to knit 4 rows of garter stitch.

Follow the pattern and keep knitting until the lace you're working with achieved your desired length.

Some consider that "true" knitted lace has pattern stitches on both the right and wrong sides, and that knitting with pattern stitches on only one side of the fabric, so that holes are separated by at least two threads, is technically not lace, but "lacy knitting", although this has no historical basis. Eyelet patterns are those in which the holes make up only a small fraction of the fabric and are isolated into clusters (e.g., little rosettes of one hole surrounded by others in a hexagon).

At the other extreme, some knitted lace is almost all holes, e.g., faggoting. Famous examples include the wedding ring shawl of Shetland knitting, a shawl so fine that it could be drawn through a wedding ring. Shetland knitted lace became extremely popular in Victorian England when Queen Victoria became a Shetland lace enthusiast. From there, knitting patterns for the shawls were printed in English women's magazines where they were copied in Iceland with single ply wool.

Knitted lace with no bound-off edges is extremely elastic, deforming easily to fit whatever it is draped on. As a consequence, knitted lace garments must be blocked or "dressed" before use, and tend to stretch over time.

Some lace is distinguished by its shape and method. Faroese lace shawls are knit bottom up with center back gusset shaping unlike a more common neck down, triangular shawl.

LACE KNITTING TECHNIQUE

A hole can be introduced into a knitted fabric by pairing a yarn over stitch with a nearby (usually adjacent) decrease. If the decrease precedes the yarn over, it typically slants right as seen from the right side (e.g., k2tog, not ssk; see knitting abbreviations). If the decrease follows the yarn over, it typically slants left as seen from the right side (e.g., k2tog tbl or ssk, not k2tog). These slants pull the fabric away from the yarn over, opening up the hole.

Pairing a yarn over with a decrease keeps the stitch count constant. Many beautiful patterns separate the yarn over and decrease stitches, e.g., k2tog, k5, yo.

Separating the yarn over from its decrease "tilts" all the intervening stitches towards the decrease. The tilt may form part of the design, e.g., mimicking the veins in a leaf.

There are few constraints on positioning the holes, so practically any picture or pattern can be outlined with holes; common motifs include leaves, rosettes, ferns and flowers. To design a simple lace motif, a knitter can draw its lines on a piece of knitting graph paper; right-slanting lines should be produced with "k2tog, yo" stitch-pairs (as seen on the right side) whereas left-slanting lines should be produced with "yo, k2tog tbl" (or, equivalently, "yo, ssk" or "yo, skp") stitch pairs (again, as seen on the right side). More sophisticated patterns will change the grain of the fabric to help the design, by separating the yarn overs and decreases.

It is common for lace knitters to insert a "lifeline", a strand of contrasting yarn threaded through stitches on the needle, at the end of every pattern repeat or after a certain number of rows. This allows the knitter to rip out a controlled number of rows if a mistake is discovered.

Internationally recognized lace expert Margaret Stove demonstrates how different combinations of yarnovers and decreases yield more solid or more airy sections of lace, and Sivia Harding takes you step-by-step through the process of designing your own unique triangle-shaped lace shawl.

Ysolda Teague explains what she loves about knitting and designing lace, and Ann Budd explains three different ways of working a perfect finish for lace by blocking. [photo Margaret Stove value study.psd; caption: Margaret Stove demonstrates how different arrangements of yarnovers and decreases create contrast in lace patterns. The issue's most mind-blowing article comes from our own Joni Coniglio, senior project editor of our knitting magazines, who was always bothered by the awkwardly visible join created by working Kitchener stitch to graft two halves of a stole. There must be a way, she reasoned, to work the graft in pattern and minimize the awkwardness of that row. To our delight (but not our surprise), she developed a surprisingly simple method of rearranging stitches with a tapestry needle, then grafting to make a nearly invisible join.

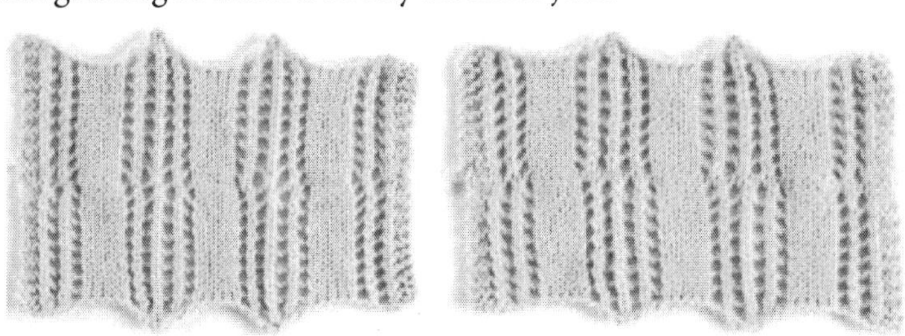

Using her innovative grafting method, Joni Coniglio avoids the noticeable seam in the left sample and creates a near-invisible seam instead.

Creativity in Every Stitch

Heather Zoppetti's Clematis Tunic

The patterns in this issue have something for all fans of lace knitting.

Flemish Block Lace

Works on multiples of 14 plus 3 stitches.

Row 1 (and all wrong side rows): Purl.

Row 2: Knit 2, *knit 2 together, yarn over, knit 1, yarn over, slip, slip, knit, knit 3, knit 2 together, yarn over, knit 4. Repeat from * across, ending knit 1.

Row 4: Knit 1, *k2tog, yo, k3, yo, ssk, k1, k2tog, yo, k4. Repeat from * across, ending k2.

Row 6: K2tog, yo, *k5, yo, slip 1, k2tog, pass slipped stitch over, yo, k4, k2tog, yo. Repeat from *, ending k1.

Row 8: K2, *yo, ssk, k4, yo, ssk, k3, k2tog, yo, k1. Repeat from *, ending k1.

Row 10: K3, *yo, ssk, k4, yo, ssk, k1, k2tog, yo, k3. Repeat from *.

Row 12: K4, *yo, ssk, k4, yo, k3tog, yo, k5. Repeat from *, ending last repeat k4.

Repeat these rows for pattern.

Many people consider knitting too three-dimensional to produce the traditionally fine, flat, and fluid fabric of lace. While this is somewhat true, the same innate depth and texture of knitted fabrics can be manipulated to produce beautiful lace effects.

Finished knitted lace may look brain-staggeringly complex. But lace is essentially composed of two simple elements: increases and decreases.

The Standard Increase: YO

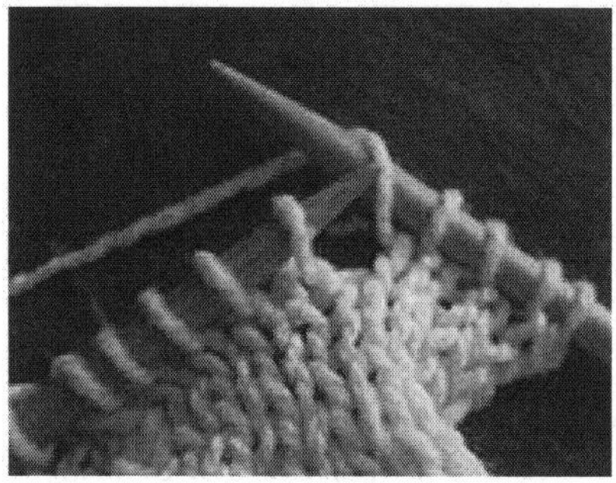

The most common increase you'll encounter in lace patterns is an open increase called a yarn-over (referred to as a YO in patterns and shown at left). The increase is made by wrapping your yarn once around the right needle, as if you were working a regular stitch but without doing anything to the stitches on your left needle. It's the simplest possible increase you can do.

By not connecting with stitches from the previous row, yarn-overs produce the large, open holes in your fabric that are essential for the open, airy look of lace. For this same reason, yarn-overs are often used sparingly and only for effect in non-lace patterns.

Decreases

Unless you're trying to create a progressively wider piece of fabric, you'll want to balance out all those increases with a reduction in stitches elsewhere in the row. Here's where the second element of lace knitting comes in: the decrease.

Right-Slanting Decrease: The K2Tog

The first decrease most knitters learn is simply knitting two stitches together, or K2tog (shown at left).

This produces one stitch that tilts to the right. The same is true if you work the decrease on a purl row, or P2tog.

Left-Slanting Decrease: The PSSO

If you're working any type of garment that needs parallel decreases tilting in opposite directions—such as working the toe decreases in a sock—you'll want to pair the K2togs on one side with another type of decrease that will cause the stitches to slant in the opposite direction, or left.

The most common way to create left-tilting decreases is to slip one stitch to your right needle without working it, knit the next stitch, and then with your left needle lift the unworked stitch over the one you just knit, letting it drop off the needle.

You'll see this written as psso, or pso, with whatever slip and knit combination noted beforehand. For example, s1k1 psso would mean slip one stitch, knit one stitch, and then pass the slipped stitch over the knit stitch and off the right needle. But you'll also see larger decreases like s1k2 to gpsso.

Here's the slipped stitch in action.

Slip one stitch to the right needle. Many knitters feel that the best way to slip this stitch is knitwise as if you were going to knit it. This will position it so that, when slipped, it has a subtle twist that more closely matches the standard K2tog look. (Ultimately, I'd say do what you like best.) Then knit the next stitch as you would normally.

Select the slipped stitch with your left needle and...

...lift it over the knitted stitch and off the needle. See how it tilts to the left?

Left-Slanting Decrease: The SSK

Recently designers have favored another type of left-slanted decrease called the SSK, or slip-slip-knit.

To do this, you begin by slipping two stitches to the right needle, one at a time, as if to knit. (Some prefer to slip the first stitch as if to knit, then the second as if to purl. Again, try all techniques and choose the one you like best.)

Then slip the tip of your left needle, left to right, behind the fronts of those two stitches, keeping the tip of the left needle towards you. Take your yarn and wrap it around the working needle to make a stitch.

Pull that loop through the two slipped stitches and let them slip off the needle. Again, see how the stitches tilt to the left?

Slipping Knitwise or Purlwise?

The real goal is to produce a left-slanting decrease that matches the right-slanting decrease most closely.

The Art of Placement

The soul of knitted lace lies in the relationship between your increases and decreases. The possibilities are surprisingly infinite.

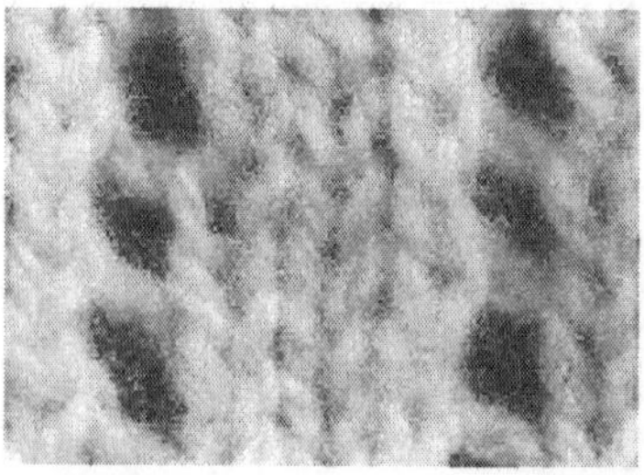

A YO followed immediately by a decrease will create a standard eyelet hole. If you follow the exact same YO/decrease pattern from row to row, the holes will stack on top of one another, as you see at left.

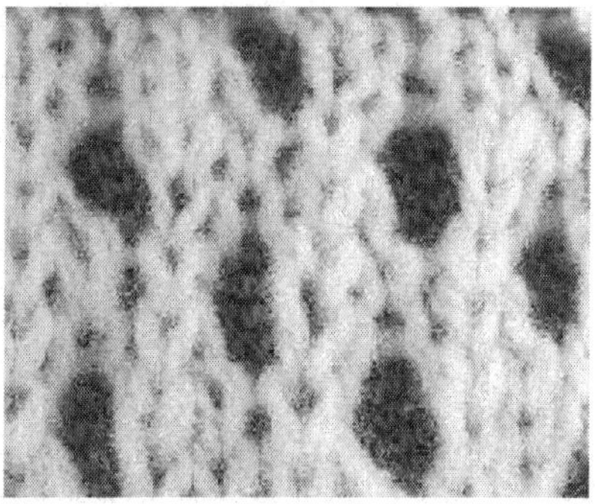

But if you stagger the repeats by a few stitches every other row (for example K2Tog, YO, K2 one one row and then K2, K2Tog, YO on the next) you can create a very simple diamond-like effect, which you see at left.

The type of decrease you use (the right-pointing K2Tog or left-pointing SKPSSO) and the location of the decrease in relation to the increase (both within individual rows and the entire fabric) will produce dramatically varied results. Designers essentially use the diagonal lines of decreases and circles of YOs to draw patterns in the knitted fabric.

The more you experiment with different patterns, the clearer this will become. For now, let's get you started on a project that demonstrates how increases and decreases can create lacy patterns.

Beginner Project: Easy Lace Scarf

Believe it or not, the elegant pattern you see at left is based on an extremely easy combination of stitches.

Perfect for beginners, this stitch—called "Feather and Fan" or "Old Shale"—will familiarize you with the basic dance of increases and decreases in lace, producing gorgeous effects that knit up quickly and easily. This stitch achieves its wavy, puckered effect by grouping all the decreases together, and then following them with a series of increases.

Feather and Fan looks particularly dramatic with finer, lace-weight yarns, but you can use a heavier yarn (such as DK) without losing too much of the intended effect.

Materials

Choose a yarn that will wear comfortably against your bare neck. I used Classic Elite Lush, a smooth four-ply blend of angora and wool that knits up at 4.5 sts per inch on US 8 needles. One 123-yard skein will produce the scarf you see here, which measures 4 x 45 inches.

The yarn's recommended needle size is US 8, but I wanted the stitches to have plenty of room so I bumped my needle size up to US 9. Fortunately you don't need to be too obsessed with gauge since this will only wrap around your neck.

Cast on 22 stitches and knit 2 rows.

Now you're ready to begin the lace pattern, which is as follows.

Row 1: (RS): K all sts.

Row 2: K2, p18, k2. (Those 2 border stitches on either end will always be knit in garter stitch—their role is to keep your fabric from curling.)

Row 3: K2, (K2Tog) 3 times, (YO,K1) 6 times, (K2Tog) 3 times, k2.

Row 4: K all sts.

Repeat rows 1 through 4 until you've reached your desired scarf length or run out of yarn, whichever comes first. End by completing Row 4, and then knit 1 more row and bind off all sts.

Finishing: Darn in any ends and fill a sink with lukewarm water and add a gentle detergent (such as Eucalan orSoak). Drop your scarf into the water and gently squeeze it until the water has fully saturated the fibers. If using Eucalan or Soak, you can skip any rinse.

Otherwise, gather the scarf together, lift it out of the sink, drain the water, refill it with lukewarm water, lower the scarf back in the water, and gently squeeze. Repeat this process until the rinse water runs clear. Then blot the scarf dry in a towel and lay it out to shape. If you're using a finer, lace-weight yarn you'll want to use pins to help open up the fabric to its desired shape.

Note: The knit stitches on row 4 (a wrong-side row) produce a garter stitch-like effect. If you'd rather have a totally smooth surface for your pattern, you can opt to purl that row (as shown in the picture at left). If you do so, prepare for extra blocking to keep your scarf from curling in on itself.

If you want to tackle this project in a finer gauge yarn such as Lorna's Laces Helen's Lace, by all means go for it. You'll just need to cast on any multiple of 18 stitches and add 4 stitches for the border.

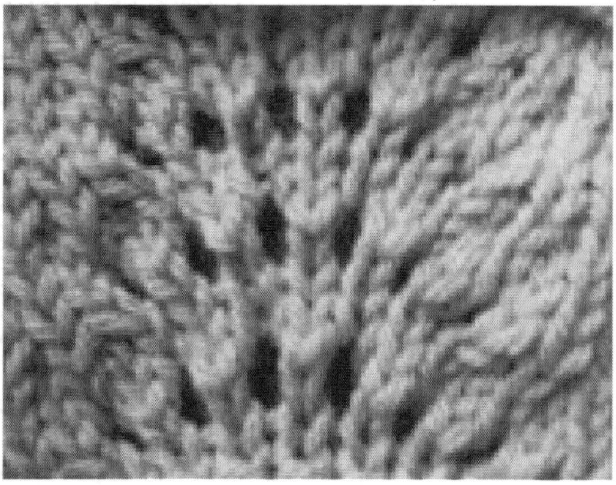

Simple examples

A horizontal row of holes can be produced by the pattern: *k3, k2tog, yo, k3*.

A pair of vertical columns can be produced by stacking the pattern: (k, k2tog, yo, k, yo, k2tog tbl, k) on the right side. Here the flanking decreases slant outwards away from the central stitch. For a thicker central column, one can move the decreases so that they slant inwards: (k, yo, dec 2 symmetrically, yo, k). For making the same pattern on the wrong side, the converse stitch patterns are: (p, p2tog, yo, p, yo, p2tog tbl, p) and (p, yo, dec 2 symmetrically, yo, p), respectively.

A diagonal row of holes can be made by shifting the (yo, dec) every row or every other row, e.g.,

- Row 1: k, k2tog, yo, k5
- Row 3: k3, k2tog, yo, k3
- Row 5: k5, k2tog, yo, k1

Lace knitting is generally not as fine as other forms of lace, such as needle lace or bobbin lace. However, it is better suited for garments, being softer and much faster to produce.

Woven fabric

A woven is a cloth formed by weaving. It only stretches in the bias directions (between the warp and weft directions), unless the threads are elastic. Woven cloth usually frays at the edges, unless measures are taken to counter this, such as the use of pinking shears or hemming.

Woven fabrics are worked on a big loom and made of many threads woven on a warp and a weft. It is stretchy only along the bias.

BASIC KNITTED FABRICS

A few types of knitted fabric are so fundamental that they have been adopted as part of the language of knitting, similar to techniques such as yarn over or decrease. Examples include stockinette stitch, reverse stockinette stitch, garter stitch, seed stitch,faggoting, and tricot. In some cases, these fabrics appear differently on the right side (i.e., as seen when making the stitch) than on the wrong side (i.e., as seen from the other side, when the work is turned).

Stockinette stitch front

Back of stockinette stitch, also same appearance as reverse stockinette stitch

Stockinette stitch (in the UK, Stocking Stitch) is the most basic knitted fabric; every stitch (as seen from the right side) is a knit stitch. In the round, stockinette stitch is produced by knitting every stitch; by contrast, in the flat, stockinette stitch is produced by knitting and purling alternate rows.

Stockinette-stitch fabric is very smooth and each column ("wale") resembles a stacked set of "V"'s. It has a strong tendency to curl horizontally and vertically because of theasymmetry of its faces.

Reverse stockinette stitch is produced in the same way as stockinette, except that the purl stitches are done on the right side and the knit stitches on the wrong side. In the round, reverse stockinette stitch is produced by purling every stitch.

Seed stitch

Seed stitch is the most basic form of a basketweave pattern; knit and purl stitches alternate in every column ("wale") and every row ("course"). In other words, every knit stitch is flanked on all four sides (left and right, top and bottom) by purl stitches, and vice versa. The Moss Stitch is another name for the seed stitch.

Seed-stitch fabrics lie flat; the symmetry of their two faces prevents them from curling to one side or the other. Hence, it makes an excellent choice for edging, e.g., the central edges of a cardigan. However, seed stitch is "nubbly", not nearly as smooth as stockinette.

Faggoting

Faggoting is a variation of lace knitting, in which every stitch is a yarn over or a decrease. There are several types of faggoting, but all are an extremely open lace similar to netting. Like most lace fabrics, faggoting has little structural strength and deforms easily, so it has little tendency to curl despite being asymmetrical. Faggoting is stretchy and open, and most faggoting stitches look the same on both sides, making them ideal for garments like lacy scarves or stockings.

Tricot knitting

Tricot is a special case of warp knitting, in which the yarn zigzags vertically, following a single column ("wale") of knitting, rather than a single row ("course"), as is customary. Tricot and its relatives are very resistant to runs, and are commonly used in lingerie.

Other basic fabrics

Other classes of basic knitted fabrics include ribbing, welting and cables.

Ribbing (knitting)

Ribbing is a border that is commonly used on the cuffs of sweaters. It is used because it produces a stretchy fabric. It is made with a combination of knit stitches and purl stitches. Ribbing can be done with any number of stitchs, but two or three is common. In our example, we use two.

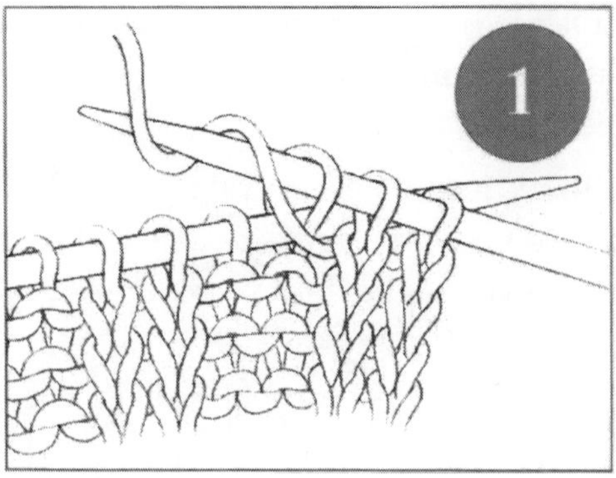

Knit two stitches. Bring the yarn from the back of your work between the needles to the front. Now purl two stitches. Bring the yarn to the back of your work between the needles. Knit two stitches. Continue to knit two stitches, purl two stitches until you reach the end of the row or until the pattern instructs you to do something different.

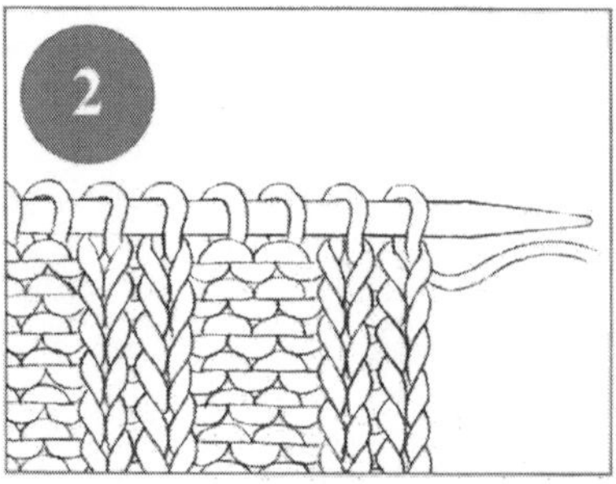

On subsequent rows, knit the knit stitches (V stitches) and purl the purl stitches (bump stitches) Remember to have the working yarn in BACK when you knit and in FRONT when you purl. In knitting, ribbing is a pattern in which vertical stripes of stockinette stitch alternate with vertical stripes of reverse stockinette stitch. These two types of stripes may be separated by other stripes in which knit and purl stitches alternate vertically; such plissé stripes add width and depth to ribbing but not more elasticity.

The number of knit and purl stripes (wales) are generally equal, although they need not be. When they are equal, the fabric has no tendency to curl, unlike stockinette stitch. Such ribbing looks the same on both sides and is useful for garments such as scarves.

Ribbing is notated by (number of knit stitches) x (number of purl stitches). Thus, 1x1 ribbing has one knit stitch, followed by one purl stitch, followed by one knit stitch, and so on.

Ribbing has a strong tendency to contract laterally, forming small pleats in which the purl stitches recede and the knit stitches come forward. Thus, ribbing is often used for cuffs, sweater hems and, more generally, any edge that should be form-fitting. The elasticity depends on the number of knit/purl transitions; 1x1 ribbing is more elastic than 2x2 ribbing, etc. However, some cable patterns may "pull in" more than ribbing (i.e., have a smaller gauge); in such cases, a ribbed border may flare out instead of contracting.

Slip stitches may be added to increase the depth of the ribbing, and to accentuate the stitches of certain wales. For example, the knit stitches can be slipped every other row to double their height and make them come forward.

Ribs can be decorated with nearly any motif used for a plain knitted fabric, e.g., bobbles, cables, lace, various colors, and so on.

Slip-stitch knitting

Slip-stitch knitting is a family of knitting techniques that use slip stitches to make multiple fabrics simultaneously, to make extra-long stitches, and/or to carry over colors from an earlier row.

In the basic slip stitch the stitch is passed from the left needle to the right needle without being knitted. The yarn may be passed invisibly behind the slipped stitch (wyib="with yarn in back") or in front of the slipped stitch (wyif="with yarn in front"), where it produces a small horizontal "bar".

The wyif slipped stitch is less common, although several knitting patterns use it to produce a visual effect like woven cloth. Alternatively, the yarn can be "tucked", i.e., made into a yarn-over that is knitted together with the slipped stitch on the next row; like the simpler wyib, this is invisible.

If knitted on the next row, the wyib slipped stitch is twice as tall as its neighboring stitches. A vertical column of such "double-height" stitches is a nice accent, e.g., on a scarf or in a sweater, particularly in a contrasting color.

DOUBLE KNITTING WITH SLIP STITCHES

Double knitting is a very easy technique to master, even though is sounds difficult and looks like magic. You use two straight needles to knit back and forth, but you end up with a tube of knitting. It's a great technique to make a double thick fabric for a yoga mat or rug, and it's also a cool way to make a pillow, hat, or bag. I used the technique in the condom amulets because I didn't want to use double-pointed needles for such a small project and I was too lazy to sew seams.

1. Cast on an even number of stitches.
2. Work every row as follows: *Knit 1 st, slip 1 st as if to purl with yarn in front. Rep from * to end of row. Turn.

That's it! Every two rows across makes one round of knitting. You are basically knitting half of the stitches on each pass. Be very careful to put the yarn in front when you slip the stitches, or you won't have an open tube. The yarn has to be in between the two layers of fabric when you slip the stitches.

I bet you thought this would be harder, didn't you?

If you lose count, it's very easy to find your place if you're using plain yarn (don't try this technique with furry yarn until you practice on smooth wool or cotton first).

Knit the first stitch and the stitches that look like Vs. These are the knits on the RS of the fabric:

Slip the 2nd stitch and the stitches that look like bumps. These stitches are hidden between the knit stitches after the first couple of rows:

3. To cast off:

If you want a double thick layer of fabric that doesn't open on the end:

- Bind off normally.

If you want an open tube for a hat or a bag or for a pillow that you'll stuff:

- Either slip the front sts to one double-pointed needle and the back sts to a second, or just pull the needle out and rearrange the stitches so the front stitches are on one needle and the back sts on a second.
- Using a thrid needle, bind off the sts on the front needle, turn.
- Bind off the sts on the back needle.

Slip stitches may be used for an easier method of double knitting that requires only one yarn be handled at one time. As a concrete example, consider a two-color pattern with a multiple of four stitches (labeled ABCD) being knit on double-pointed circular needles.

On the first row, using color 1, stitch A is knitted, stitch B is purled, stitch C is slipped wyib and stitch D is slipped wyif. The knitter then slides the stitches back to the beginning (recall that the needles are double-pointed). Then, using color 2, stitch A is slipped wyib, stitch B is slipped wyif, stitch C is knitted and stitch D is purled. The knitter then turns the work and repeats indefinitely.

The knitted and slipped wyib stitches come forward, whereas the purled and wyif stitches recede, resulting in a (very warm!) double-knit scarf alternating in the two colors with beautiful drape. The knit and purl stitches produce the front and back fabrics, respectively, of the double-knitted fabric while the slipped stitches allow for the alternation of color.

An even simpler slip-stitch pattern generates two fabrics at once on the same needle. Consider the pattern: * knit 1, slip 1 wyif *. At the end of the row, turn the work. Then knit the stitches that were slipped and slip (again wyif) the stitches that were knitted. In the end, one should obtain a "pocket" that can be opened (be sure to use wyif slip-stitches during binding off as well!) The wyif slip stitch prevents the yarn from crossing over to the back fabric, so that only the front fabric is knitted in any row. This is probably the secret technique of Anna Makarovna from Leo Tolstoy's War and Peace, who always knit two socks simultaneously When the pair was finished, she made a solemn ceremony of pulling one stocking out of the other in the presence of the children.

MOSAIC KNITTING

If you've stuck to knitting in single colors, or maybe just ventured into stripes, here's some good news: You can achieve what appear to be complicated color patterns while using only one yarn at a time! Many knitters suffer from fear of Fair Isle because they are loathe to carry two colors at once. It seems complicated and unwieldy. If you're ready for a little more adventurous use of color and pattern, why not try mosaic knitting? Using a simple slip-stitch technique, you can create bands and borders in pattern, or even a whole sweater, and you won't ever have to carry a second yarn along!

Color slip-stitch knitting has been around for a long time, but it was brought into the spotlight for a time by Barbara Walker who dubbed the style "mosaic knitting" in the late '60s. Hundreds of mosaic patterns appear in Walker's books. More recently, Roxana Bartlett has taken a new look at this approach to color patterning. If you become smitten with this technique after trying it, you can find more information and patterns in their books. [See "Recommended Reading" below.]

The technique of two-color slip-stitch knitting is really quite simple: If you can knit simple stripes and slip a stitch, you have all the skills you need. Mosaic knitting simply involves slipping the stitches in a row that should be the "other" color. If you are knitting the dark color, you slip the light; if you are knitting the light color, you slip the dark.

Aside from the allure of color patterning using just one yarn at a time, slip-stitching has other advantages over other multicolor techniques: Mosaic knitting works well either flat or in the round which gives it more flexibility than intarsia or Fair Isle techniques. It also doesn't leave as many floats as Fair Isle knitting, so the fabric is not as heavy and not as much yarn is used.

There are a few limitations to the technique. Because stitches are slipped, there's a limit to how many stitches of the "wrong" color you can work consecutively — two or three, depending on who you talk to. This tends to give the patterns created a geometric look, which may have reminded Barbara Walker of tile mosaics. Also, the slipped stitches make the fabric a little less than smooth, but blocking will usually take care of this. For this reason, yarns with a little more "give" tend to work better, though the toddler sweater above is knit in pure cotton. If you knit very tightly, you may want to go up a needle size so that your stitches can be slipped two rows without puckering.

I find slip-stitch knitting to be a great introduction to color patterning. There are bits of "Pay attention; read the chart!" combined with the auto-pilot knitting of every other row. And once you get the feel of a certain motif, you will be able to see it in your knitting without the chart.

READING the chart

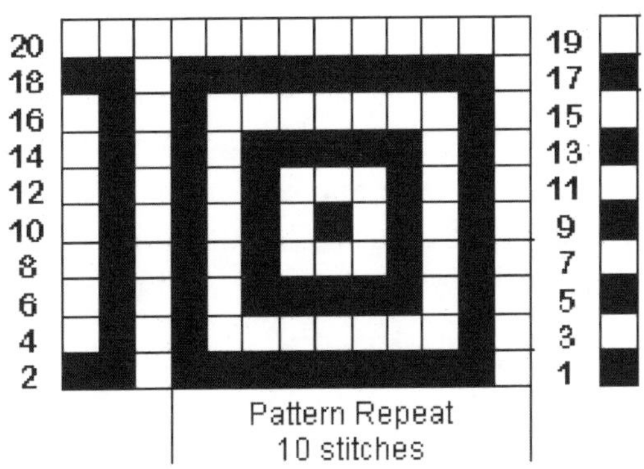

Usually mosaic patterns are charted; most are not very large [maybe 20 stitches wide], but there is really no limit. The charts usually have a color bar on the right side indicating which color is being worked for each row. The chart has the rows numbered, odd numbered rows on the right, even numbered rows on the left. The second row worked in a color is always worked exactly the same as the first: if a stitch was knitted in the odd row, it is worked in the even row; if a stitch was slipped in the odd row, it is slipped again.

Sometimes, particularly for simpler motifs such as the brick pattern at right, directions are written out like standard knitting patterns.

MC: CO a multiple of 6 sts -1.

Row 1: Knit.

Row 2: Purl.

Row 3: CC: *K5, sl1* end k5.

Row 4: *P5, sl1* end p5

Repeat these 4 rows as until desired length is reached.

Getting started

After choosing a mosaic pattern [or patterns] you like, you can easily map it on to any garment. While you want to repeat the entire motif to see the whole pattern, you need not necessarily have a number of stitches that is an exact mulitple of the number of stitches in the motif. You may cut a motif off in the middle without making a mess. With any project, you will want to work the selvedge stitches in the row's color. Never slip the selvedge stitch. This will keep your pieces smooth and easy to join together.

After casting on, work two rows in the first color [say, the lighter color], then attach the second color, leaving the first hanging. With the darker yarn, the light-colored stitches shown on the chart are slipped and the darker stitches are worked. Always slip stitches as if to purl; that is, insert the needle from right to left, and carry the yarn on the wrong side of the work. After you have knit your second row with the darker color, switch back to the light. You will always work two rows in each color and the two rows are the same: a stitch never changes its color in the second row. Continue working your way up the chart until you complete the motif.

After you have finished with one motif, you may repeat it again, either the same way, or by trading light and dark colors; switch to a second motif, or carry on with plain knitting. A mosaic border on cuffs and hem will add punch to an otherwise plain sweater, or try working a stripe of a larger motif across the chest of a sweater. And though I have described these techniques as two-color, that is certainly not the limit of this technique. You can keep one color constant and alternate the second. Or try one plain yarn and one variegated yarn. Using a plain colored yarn and one space-dyed yarn can yield very impressive multicolor results - all just one strand at a time.

Mosaic knitting uses two colors (usually both held at one side), but only one yarn is handled at one time. Let the first and second yarns be called "black" and "white" for specificity, although any two colors may be used. The knitter casts on an entirely white row. The knitter then introduces the black yarn and knits two rows, across and back.

If a white stitch is desired at a given position, the stitch (from the white row below) is slipped wyib; by contrast, if a black stitch is desired, the stitch is knitted using the black yarn. The second (return) row repeats the slip/knit choices of the first row, effectively giving double height. (Note that, on the return row, the stitches are slipped wyif, since the fabric is being knitted from the wrong side.) The knitter then takes up the white yarn and knits two rows, across and back.

Now if a white stitch is desired at a given position, the stitch is knitted with the white yarn; by contrast, if a black stitch is desired, the stitch is slipped from the row below (if it is black). If a black stitch is needed in a white-yarn row and the stitch of the previous row was white (i.e., slipped), the pattern is impossible for mosaic knitting. Therefore, any black or white vertical stripe must begin and end with the corresponding yarn, which implies that the number of knitted rows in any vertical stripe must be 2 times an odd number, i.e., 2x1=2, 2x3=6, 2x5=10, etc. However, this constraint on possible patterns can be well-hidden if the pattern is large enough.

Mosaic knitting can produce many beautiful patterns, particularly geometrical or Grecian designs. Historically, mosaic patterns are rather rectilinear, being composed mainly of thin horizontal and vertical stripes that meet at right angles.

However, mosaic knitting has limitations relative to other techniques for producing color patterns in knitting such as Fair-isle knitting. Depending on the pattern, a mosaic-knit fabric may be stiff and tense, due to the many slipped stitches; such fabrics may be better for coats and jackets, which do not require as much drape. The tension in the fabric may also distort the rectilinear lines into curves.

These problems may be overcome by judiciously elongating the stitches. Some color patterns may be impossible for mosaic knitting, if they require too many slipped stitches or if the colors do not line up conveniently (as described above). Blocks of solid colors can be done in mosaic knitting, but require many slipped stitches in a row; hence, blocks of solid colors are usually broken up with stippling, i.e., with regularly spaces spots of the opposing color.

On the other hand, mosaic knitting is significantly lighter than Fair-isle knitting, which is nearly twice as thick and bulky. Long horizontal bars of the same color are also more convenient in mosaic knitting than they are in Fair-isle knitting (where such bars require that the other colors be held at the back for long runs).

A simple extension of mosaic knitting is to use multiple colors, rather than just two, although the limitations on the pattern become even more severe. Another variant is to hold the yarns on opposite edges (or to knit circularly), which allows the knitter to knit only one row per color.

Basic knitted fabrics

Knitting is the construction of an elastic, porous fabric, created by interlocking yarns by means of needles. Knitted fabrics can be made much more quickly and easily than woven fabrics at comparatively less cost. Knitted fabrics are generally light in weight, comfortable in wear even during travel, but yet require little care to keep their neat appearance. The tendency of knits to resist wrinkling is another factor to boost up their popularity. Knitted fabrics are used for designing active clothing such as sports clothing. Their elastic nature permits for abundant physical activity. Knitted fabrics are produced by two general methods. Warp knitting and weft knitting. They are made as flat or tubular fabrics depending on the end use. Tubular fabrics may not have any seams at the sides where as flat fabrics are treted just like woven fabrics.

KNITTED FABRIC TYPES

Jersey Knit

These fabrics are weft knitted and are characterised by distinct but flat vertical lines on the face and dominant horizontal ribs on the reverse side. Fancy varieties are also produced. They are used in making hoisery sweaters, sports wear etc.

Rib knits

These fabrics are made by using rib stitch with two sets of needles. These fabrics are used where stretch is desired as they show excellent degree of elasticity. Rib knits are warm to wear. They are used as apparels such as shirts, blouses, body stockings etc.

Double knit

Double knits are produced by the interlock stitch. The fabrics is riblike in appearance on both the sides. Decorative fabrics are also produced by jacquard attachment. These fabrics show good dimensional stability and are easy to cut and sew. They do not require any seam finishes as the fabric does not ravel. They are firm, heavier, less stretchable and more resilient. Double knits are commonly made from polyester, cetate or wool fibres. They find use as every durable apparels.

Knitted fur fabrics

A wide range of knitted fur fabrics are available in the market. The fabrics are produced by pile knitting. The extra set of soft filament yarns used form the pile on the surface of the fabric. The pile is cut and the fabric are finished similar to the original fur. These fabrics are mainly used for coats and trimmings for other dresses.

A few types of knitted fabric are so fundamental that they have been adopted as part of the language of knitting, similar to techniques such as yarn over or decrease.

Examples include stockinette stitch, reverse stockinette stitch, garter stitch, seed stitch, faggoting, and tricot.

In some cases, these fabrics appear differently on the right side (i.e., as seen when making the stitch) than on the wrong side (i.e., as seen from the other side, when the work is turned).

Stockinette stitch and reverse stockinette stitch

Stockinette stitch front

Back of stockinette stitch, also same appearance as reverse stockinette stitch

Stockinette stitch (in the UK, Stocking Stitch) is the most basic knitted fabric; every stitch (as seen from the right side) is a knit stitch. In the round, stockinette stitch is produced by knitting every stitch; by contrast, in the flat, stockinette stitch is produced by knitting and purling alternate rows.

Stockinette-stitch fabric is very smooth and each column ("wale") resembles a stacked set of "V""s. It has a strong tendency to curl horizontally and vertically because of theasymmetry of its faces.

Reverse stockinette stitch is produced in the same way as stockinette, except that the purl stitches are done on the right side and the knit stitches on the wrong side. In the round, reverse stockinette stitch is produced by purling every stitch.

Seed stitch

Seed stitch is the most basic form of a basketweave pattern; knit and purl stitches alternate in every column ("wale") and every row ("course"). In other words, every knit stitch is flanked on all four sides (left and right, top and bottom) by purl stitches, and vice versa. The Moss Stitch is another name for the seed stitch.

Seed-stitch fabrics lie flat; the symmetry of their two faces prevents them from curling to one side or the other. Hence, it makes an excellent choice for edging, e.g., the central edges of a cardigan. However, seed stitch is "nubbly", not nearly as smooth as stockinette.

Faggoting

People always say the 80's are here and are never going away lamenting the continuous return of retro style, this kind of observation is akin to saying that Asian food is in is and never going away. Do they mean Thai? Do they mean Korean, Vietnamese, or Indonesian? Sushi? The spectrum of culinary opportunities across the Pacific and Indian oceans are just as wide as the hodgepodge style dictum that made up the 80's. For every look there was a counter look, for every culture there was a counter-culture, the possibility for contemporary rejuvenation is endless. With the current vulgarity in mainstream women's fashion is it any surprise that designers are finding these particular inspirations relevant?

Faggoting is a variation of lace knitting, in which every stitch is a yarn over or a decrease. There are several types of faggoting, but all are an extremely open lace similar to netting. Like most lace fabrics, faggoting has little structural strength and deforms easily, so it has little tendency to curl despite being asymmetrical. Faggoting is stretchy and open, and most faggoting stitches look the same on both sides, making them ideal for garments like lacy scarves or stockings.

There's no institution more fascinating than that of a fashion brand. Designers and the houses they design for, their own names or not, strive to create or uphold their own institution with every collection they present. How do you handle the burden of a legacy, spin it into something new, build one into something covetable, or simply maintain the potency of past grandeur?

Rei has built her company, Comme De Garcons, on the premise that fashion is a medium that can nurture and also demand a reactionary voice. She shocked the 80's and the stuffy fashion establishment of the time and this cause has been the theme of her label ever since.

But since then we've seen so many alternatives come and go, it brings into question the relevance of her ant-fashion/anti-pretty rigor. For a generation that regards deconstruction, full black ensembles, and uncomfortable allusions old hat what else does she have to say? Does it go beyond a clatter of color and technique?

Does it resonate beyond making something challenging and creative for the selective few that will appreciate it? I went into the CDG store in Chelsea recently while waiting for a friend at a nearbye bookfair. It's appropriately settled amongst the notorious galleries, the whole area bleeds a concious effort of art. But like the institution Chelsea has become the clothes on the racks seemed just as stagnant as the art (in all it's good intentions) across the street.

And then you have the stuffiest and most moribund institution in fashion, Yves St. Laurent. St. Laurent quickly built his name into a juggernaut force. And as time went on his own acuteness waned. It's fair to say that St. Laurent was not at the top of his game the last 10-20 years of his career. Of course it's neither here nor there, my point is not to discredit his greatness. It has more to do with the conundrum that Stefano Pilati inherited. What is he to do?

His challenge is simple but strenous, to resubmit the creative force and chic aura that the young St.Laurent triumphantly crafted, and he must sell. He's gotten a hard time with the editors, some expecting too much or simply not evaluating him on the right merits.

But I find that when Pilati trusts his own instincts the most is when the nexus of his own talents and that of St. Laurent becomes so engorged and bares the most sincerity. Pilati has proven himself over and over and this most recent collection is perhaps his finest. But does it make any money? It's a "but" that stings and I'm sure one he's infinitely aware of. How does he fill the footsteps of a giant? Perhaps he can't, but he's certainly making his own.

TRICOT KNITTING

Tricot is a special case of warp knitting, in which the yarn zigzags vertically, following a single column ("wale") of knitting, rather than a single row ("course"), as is customary. Tricot and its relatives are very resistant to runs, and are commonly used in lingerie.

Other basic fabrics

Other classes of basic knitted fabrics include ribbing, welting and cables.

Step 1: Casting On

If possible, cast all your stitches onto one needle. You can put a tip protector or rubber band on the non-working end of the needle to keep stitches from falling off.

The reason for this is simple: You want your cast-on stitches to be as consistent in gauge and appearance as possible. Casting on multiple needles introduces the risk of slipped, twisted, or loose stitches at each needle join.

Step 2: Dividing Stitches onto Each Needle

Depending on which technique you use, you'll want to divide your stitches evenly onto either three or four needles. The most common technique is to work on three needles and use the fourth needle for knitting, so that's the technique I'll describe here.

The easiest way to divide your stitches is to lay the needles flat on a table. You can fix any twists and then easily slide stitches onto the DPNs without much additional jostling.

Lay the work so that the left-most needle has your first cast-on stitches and the right-most needle has the working yarn.

Step 3: Join Needles

At this point you should have all your stitches evenly divided on the needles. They should be arranged in one long row on a flat surface, with the working yarn and final cast-on stitches on the right-most needle.

Give the stitches one final check to make sure they all face the same direction and haven't twisted—something cast-on stitches love to do.

Now take the left needle in your left hand and the right needle in your right hand. Bring these two needles together, with the middle needle forming a triangle.

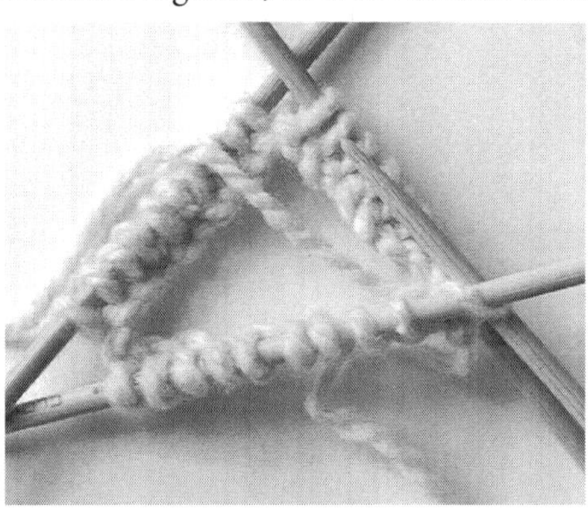

Something still seems odd, you say? That's because you haven't introduced the final, most essential element: the working needle! This is the extra needle you'll use to knit your stitches.

Step 4: Begin Knitting in the Round

With the triangle still in place (the working yarn should be on your right), take a new, empty needle and begin knitting the stitches on the left needle. In doing so, you've joined the work and are now knitting in the round.

Don't worry if there appears to be a gap in that first join. After a few rows it will tighten up, and you can fine-tune it later.

Congratulations, you did it!

Extra Tip: Beware of Ladders

As mentioned in the DPN Overview, you'll need to knit the first and last stitches of each needle more tightly than the rest. This is because stitches have a tendency to loosen at the spot where two DPNs join, which can cause long "ladders" of loose stitches.

You can also randomly move the end stitches from one needle to the other so that there's no clear join ridge. The only drawback here is that it can throw off your stitch count if you aren't careful.

If at First You Don't Succeed...

Your knitting may feel loose and precarious for the first few rounds, but don't fret. Once you get going, these stitches will stay put and you can begin to gather speed.

This whole cast-on process gets significantly easier with time, so if you have problems at first, don't give up. You'll be glad you stuck with it.

11

ADVANCED KNITTING TECHNIQUES

Advanced Knitting Technology provides complete coverage of the latest innovations and developments in knitting technology, including emerging methods as well as the latest best practice for classical processes. Many technologies can be used for the production of cloth such as weaving, knitting, nonwoven, and braiding. Knitting methods are being selected for a growing range of applications due to the spectacular properties of knitted fabric, such as softer tactile quality, higher stretchability, bulkiness, and functional properties that compare favorably with other woven fabrics.

Beyond the well-known apparel applications, specially designed knitted structures are uniquely suitable for high performance applications like reinforcement for composites, medical implants, and geotextiles.

Knitters are a lot more hesitant to call themselves "advanced" than they are to call themselves "intermediate," and that isn't surprising. Some advanced techniques are genuinely scary; take steeking, for example, which requires you to knit in the round and cut your knitting vertically. Imagine taking scissors to your precious knitting, which you just spent hours upon hours completing! It's no wonder that most knitters wouldn't touch an "advanced" pattern with a ten-foot pole.

While new techniques are being developed all the time, one shouldn't be modest and not call him or herself "advanced" just because he or she can't tackle them all. As with everything, knitters are always learning and growing, and there is no technique too difficult for someone to learn. Rather than view one of these techniques as something you can or cannot do, instead try thinking of them as skills you still get to learn. That said, I recommend becoming comfortable with the list of intermediate knitting techniques at least before attempting your first advanced pattern.

JOINING YARN

If you can locate a join at a seaming location, that is always best. You can just run the ends along the inside of the seamed edge, and it's invisible. All other joins are slightly imperfect, so best to locate them inconspicuously if possible, like under the arm of a sweater.

Felting will only work with wool, and some other animal fibers. Always try a test sample first with your yarn to see if it will work, and to see if you like the results! If you are using a smooth, shiny wool, felting will give it a dull look where felted, and a stiffer feel. For some yarns, it looks so different that the join stands out, and is not recommended.

For non-felted joins, you will have an end that pops out of work slightly; make sure this end is where you want it, by leaving a bit of yarn (1/4"-1/2") sticking out on the inside of the garment, or in as inconspicuous a place as possible.

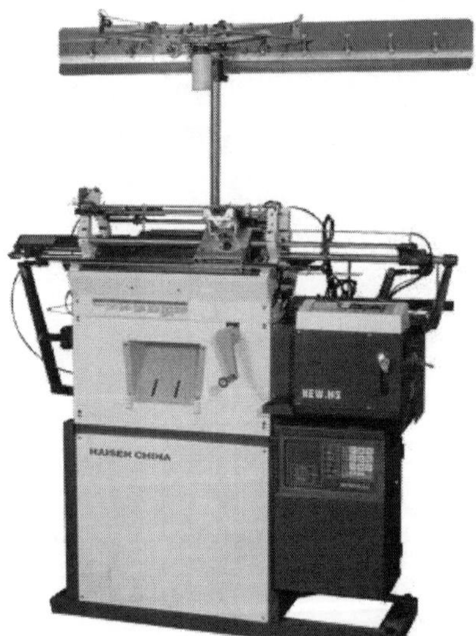

HS-308 series full computer glove machine uses the high quality steel products, integrating advanced processing equipment, guarantees products reliability, accuracy and abrasion resistance. ZHEJIANG HAISEN KNITTING MACHINE TECHNICAL CO., LTD. takes that of improving unceasingly the knitting mechanical technology to achieve the world advanced level as its important task, then exploits the more advanced knitting technology machinery, to develop our national industry vigorously.

Main specifications & parameters:

Type: HS-308

Needle pitch: 10G 13G

Size: L: 63 L: 89

Knitting speed: finger: 220r/m, palm: 200 r/m

Power: single-phase or three-phase: 220v, 250w; frequency: 50/60 Hz

Driving system: crank drive

Knitting system: sinkers knitting theory

Abnormity stop: broken yarn, poor dropping, overloading of motor, bump of needles, low battery, wrong direction

Screen display: digital display

External dimension: 1230*650*1750 (L*W*H)

Net weight: 240kg

HOW DOES A KNITTING MACHINE WORK?

- In its simplest form, a knitting machine produces flat pieces of fabric (like the sections of a sweater, pillow cover or an afghan) on a long, narrow bed of hooks that hold loops of yarn. The operator pushes a carriage threaded with more yarn across the hooks, which catch the yarn off the carriage and pull it through the loops, creating new loops. The row of knitted stitches drops down below the new loops. The operator then pulls the carriage back in the opposite direction to create another row.

David Demchuk, who writes a column for the online knitting magazine "Knitty," says the action more closely resembles weaving than hand-knitting. Knitting machines turn out fabrics with very even, meticulous rows of stitches, so they have a distinctive, machine-made look.

Standard or Bulky?

- Manufacturers make knitting machines in standard or bulky configurations. Standard machines can't handle any yarn thicker than fingering weight, which knitters use for baby clothes and very fine-knit sweaters. Bulky machines knit heavier yarns, with results that more closely resemble those of hand-knitting.
- Sponsored Links

- o Addi Click Needles
- o Addi Clicks available now Free shipping - no minimum
- o Addineedleshop.com

Speed

- The main advantage of knitting on a machine is speed. Demchuk estimates that an operator can produce between 600 and 1200 stitches per minute, and up to 20 inches of 100-stitch-wide fabric in about 5 minutes. If you've adjusted the machine properly, every stitch in the work will be perfectly even.

Drawbacks

- Still, a knitting machine has some drawbacks. For one, any stitch more complicated than a knit stitch requires the knitter to use an attachment; to manipulate the stitches on the hooks by hand; or to invest in punch cards or electronic systems to produce them. This includes the ordinary rib stitch (knit 1, purl 1 or a variation) that knitters use routinely to create the waistband and sleeve cuffs on sweaters.

Demchuk notes that many machine knitters resort to taking the work off the machine and knitting the ribbing in by hand. Another, more expensive solution to this problem is buying a double-bed knitting machine, which has one bed for making stockinette pieces, and a second for doing ribbing and lacework.

Another potential problem is that though the bed can usually accommodate at least 100 stitches, that may not be enough to make wide pieces like baby blankets or Afghans. You'll have to solve this by either buying extensions to go on the ends of the bed, or by knitting your project in pieces and then sewing them together.

Sure these techniques are impressive, but they're nothing compared to the accomplishments of learning to knit and purl. Don't be intimidated by this chapter, there are some thrilling knitting techniques in here! If nothing else, please learn how to use a circular knitting needle for large diameter circular knitting, and learn one of the techniques for small Diameter circular knitting. The techniques couldn't be easier, and we promise it will open up a whole new world of knitting.

Short Row with Wrap

This technique is used to insert extra rows invisibly in the middle of the knitting (extra rows are shown in pink in the photo, otherwise they would appear so seamless as to be invisible). The wrapping of end stitches avoids creating holes where you turn the work.

This method can be used by a knitter in many circumstances. It can be used below the neck on the back of the sweater, so the neck doesn't dip down in the back. It can be incorporated to create shaping in the chest of a sweater, great if you are a C-cup or larger, and have difficulty getting sweaters to fit in a shapely way. You can modify existing patterns by inserting rows at the underside of the chest.

One-Row Buttonhole

If the button you're using is 4 stitches wide, make a 3-stitch buttonhole (as demonstrated). For a 5-stitch-wide button, make a 4-stitch buttonhole, etc.

- Step 1 -- work to where you want buttonhole to begin. Bring yarn to front; slip 1 purl-wise, bring yarn to back.
- Step 2 -- Slip 1 purl-wise, pass first slipped stitch over second. (repeat this part 2 more times--passing over 3 stitches total, for a 3-stitch buttonhole) Place last stitch back on left needle. Turn work.
- Step 3 -- Cast on 4 stitches (for the 3 stitches you bound off, plus 1) using the cable cast-on method.
- Step 4 -- Turn work (so RS is facing). Bring yarn to back; slip first st of left needle onto right needle and pass last CO st over it.....You're done! Work to end of the row, or to the next buttonhole!

This Knit along was designed to help a knitter get started on top down knitting, and to help visualize how they are constructed.

If you are an intermediate or advanced knitter, but have never knitted a top down sweater before, these patterns might be confusing at first. This method of construction is very different from other sweater patterns.

I do not recommend knitting a top down sweater as your very first knitted garment, unless you have a patient knitter to help you get started.

To knit these patterns, as any other pattern, you will need to know how to knit, purl, increase, decrease, cast on and bind off with confidence. I suggest having a good learn to knit book on hand for a reference. You will need one of the Knitting Pure & Simple patterns to follow. Other top down patterns might be written differently.

The following photos were taken as I knitted a top down pullover for a baby.

The basic construction is the same for all of the Knitting Pure & Simple raglans. As always, you are encouraged to change edgings, body and sleeve length to customize your garment. The patterns are a perfect blank canvas for your color and stitch ideas.

For info on how to choose the size to make, and how to choose the yarn, see my frequently asked questions page.

Checking the Gauge

A very important first step! A small difference in gauge will make a large difference in finished size. Please don't make the mistake of ignoring this step.

If your sweater is to be knitted in the round, your gauge swatch will need to be knitted in the round. This is because most people purl looser than they knit, and a stockinet stitch swatch in the round has no purling, so may be tighter than one knitted back and forth.

These swatches are knitted in the round on doublepoint needles, on 34 stitches. A gauge ruler is used, to make it easier to see halves or quarters of stitches.

First photo shows a gauge of 4 1/2 to the inch, too tight! I need a larger size needle to make the knitting looser.

Every knitter is different, and lots of people need a larger or smaller needle to get the correct gauge.

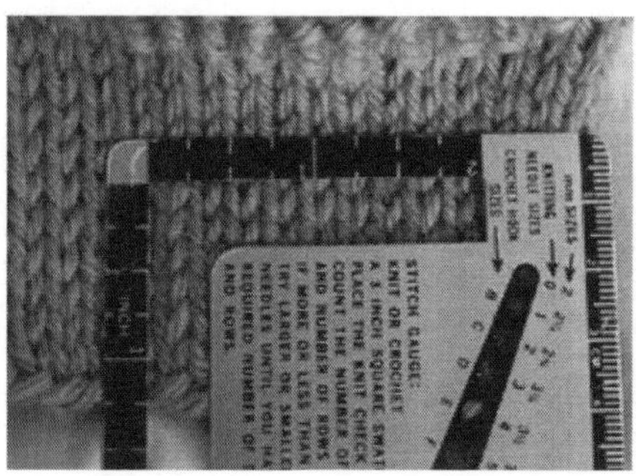

Second photo shows a gauge of 4 to the inch, which will make my sweater be the correct size. I will use the larger needle to knit the sweater.

My pattern calls for a 16 inch circular needle.

Casting on

The markers I use are made from scrap yarn. They are easier for me to use than plastic or rubber. You should use the ones that you like. The markers are placed on the needle between stitches.

In the first picture, I have started casting on using the long tail method. Use your favorite cast on, but I don't recommend the single loop cast on here. The cast on needs to be firm or slightly tight, since it will be supporting the weight of the body and sleeves. The photo below right shows how I make the markers.

The third picture shows the stitches all cast on, ready to knit row 1. You will turn the work and work back on the stitches that you just cast on.

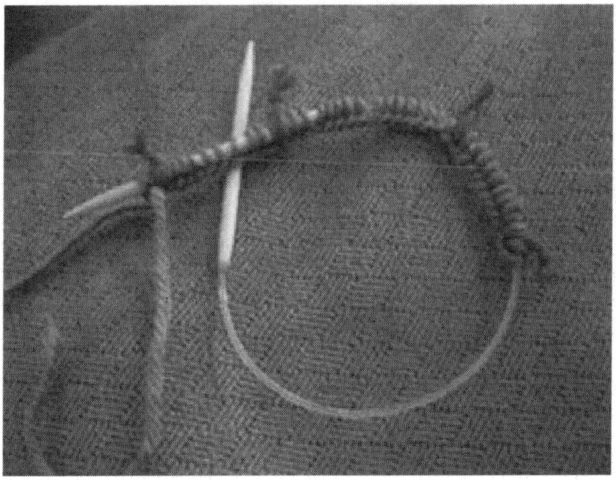

First Row

When you follow the pattern, you will be increasing in the stitch before and after each marker, and in the first and last stitch in this row.

The knit in the front and back of the stitch gives you a little purl bump, which will help you keep track of increases when you begin working in the round.

When you get to a marker, just slip it from the left needle to the right needle.

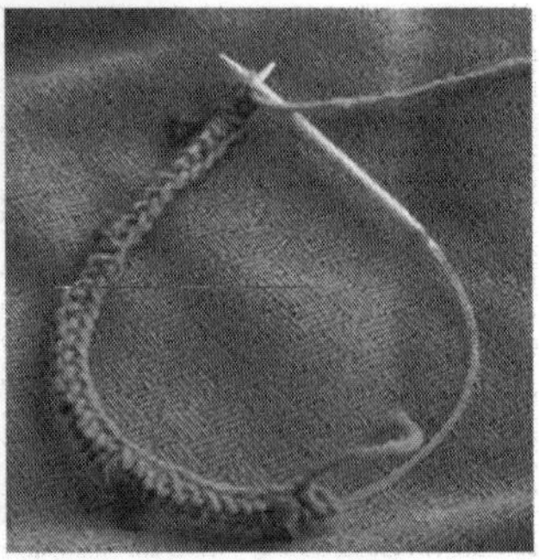

Beginning Row 2

This row is all purl, and you will slip the markers from the left needle to the right needle as you come to them.

The pattern now tells you to work one knit row with increases, and one purl row, until you have a certain number of stitches between the back markers. These are the second and third markers that you placed. In the photo above, there is a little star on the stitches you will count for the back.

At the end of the last increase row, the pattern says to cast on a certain number of sts. This will be the front of the neck opening. At the left is a photo of how you cast on these stitches.

Yes, we have changed yarns! A variegated yarn will help you see the next steps. This yarn is a vintage wool by Red Heart.

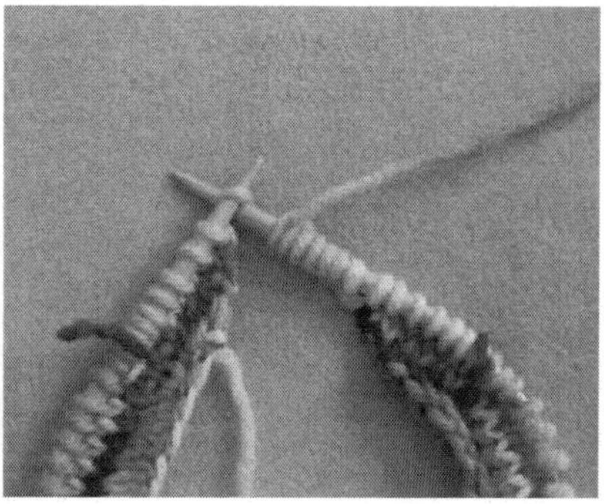

Now we will join our work into a round. To do this, just knit into the first stitch at the other end of the needle, being careful not to twist your work.

Below, your work is now in a round, with a different color marker placed to mark the beginning of the round. You are knitting one round plain, with no increases.

Your pattern now says to knit every round, working one increase round, then one plain round. Make sure that you do not increase at the pink marker, which marks the beginning of the round.

When keeping track of the increase and plain rounds, it is very easy to forget which round you are working. Making a mark on a piece of paper, or using a counter are not reliable methods of keeping track. Here is a method of keeping track that is easy and foolproof.

When knitting around, stop when there is one stitch before the next marker. Look carefully at that stitch.

Do you see that there is a little purl bump just under the needle? That purl bump is made when you do the knit in the front and back of the stitch increase. When it is just under the needle as shown, that means that you increased on the previous round. So when it looks like this, you should be doing a plain knit round. You will knit that stitch. If you look carefully, you can also see the other increases you have made in previous rows.

Now look at that same stitch. The purl bump is not right next to the needle, but a bit below. This means that your last round was a plain knit round, and that you should now be doing an increase round. So you will increase (inc) in that stitch.

Seeing your increase is easier the more you look for it. Reading your knitting this way will make what ever you knit easier and more accurate.

Dividing for the Sleeves

Your pattern now tells to work until you have a certain number of stitches between the back markers. You will then divide the sweater into sleeves and body. This is one of the most fun parts! The photo below left shows how to use darning needle and scrap yarn to hold the sleeve sts. Below right is the first sleeve on yarn holder

Hard metal holders are not recommended.

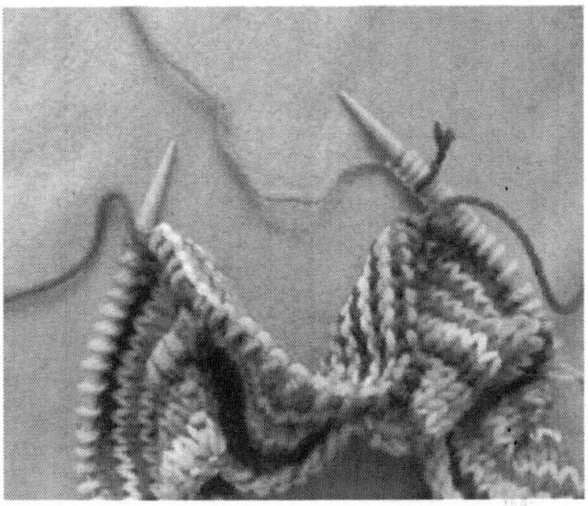

Old beginning of the round marker (pink) has been removed, red marker is new beginning marker. Yarn holders are tied and trimmed to stay out of the way.

It looks more like a sweater now! You are now working on body sts only. You will knit every round until work measures the length indicated on the pattern.

The body is almost done. Measure from top of shoulder. Our sweater will have a rolled lower edge, which is created by working several rows of plain knit on a smaller needle.

Change to smaller needles by just beginning to knit with smaller needle. Make sure you bind off with a larger needle to make your bottom edge loose. If it curls up too much, don't worry. This will be corrected in blocking. Blocking is described at the end of this tutorial.

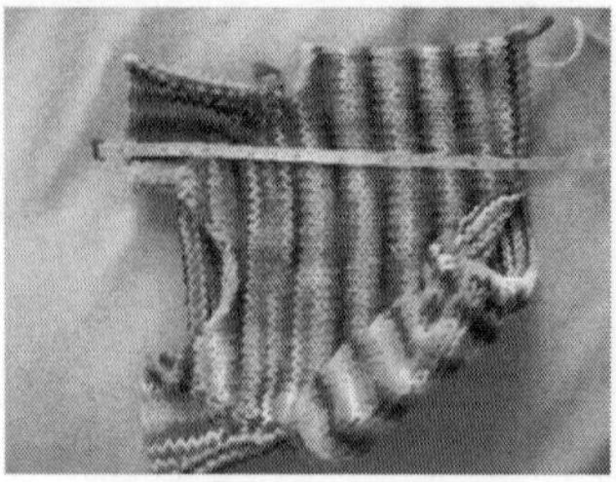

Body finished, we are now beginning the neckband. The neckband is done before the sleeves so that if you need to try on the sweater to decide how long the sleeves should be, the neckband will be there to make the sleeve length accurate.

Below left is a close up of how to pick up stitches. This is an important part of your knitting. It needs to be done neatly and carefully. Sloppy picking up can spoil the look of your sweater!

Insert the tip of the needle under the edge of the knitting, wrap your working yarn around the tip, and pull a new stitch through. Your pattern will tell you exactly how many stitches to pick up in each section. The sections are divided by the 'seam' lines created by the increasing you did.

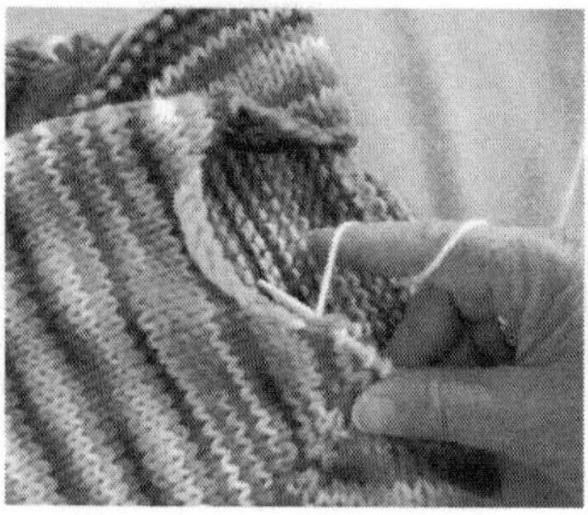

Neckband stitches all picked up, and stockinet stitch begun.

The collar is finished, and bound off very loosely. Use a size 11 needle if necessary to get this loose enough to fit over the head.

Here is a close up showing how to count rows to make sure your decreases are spaced correctly. The above decrease ratio is every 5th round. It might be hard to see at first, but with practice you will have an easy time knowing when to decrease.

One sleeve done! The other is in progress. When both sleeves are done, only weaving in ends and blocking are left to do. Sleeve length is measured from the underarm pickup.

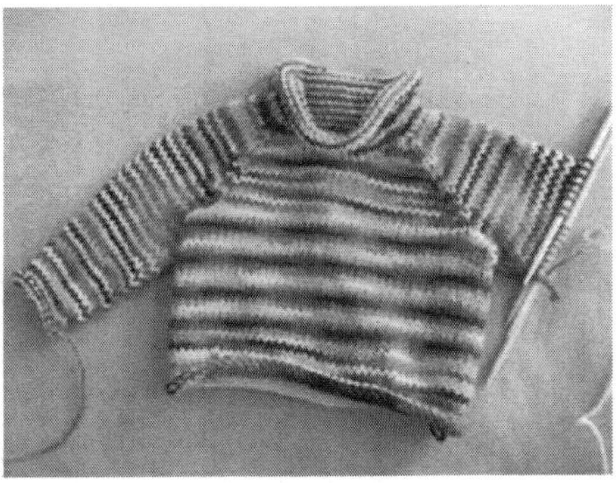

In the photo below on the left, I show how I weave in ends. I use a sharp darner to run the end in at an angle, just catching lightly the backs of the stitches, for about 2 inches.

After binding off, you will notice a little "step" where you end the round. These photos show how to even this out. This end will be woven in also.

As Lily would say, Ta Daa! The photo on the left is the sweater before blocking, and below after blocking.

To block, I wash the garment gently by hand in mild soap, rinse well, spin in the washer to get most of the water out, then lay flat to dry. The rolled edges will stay rolled this way.

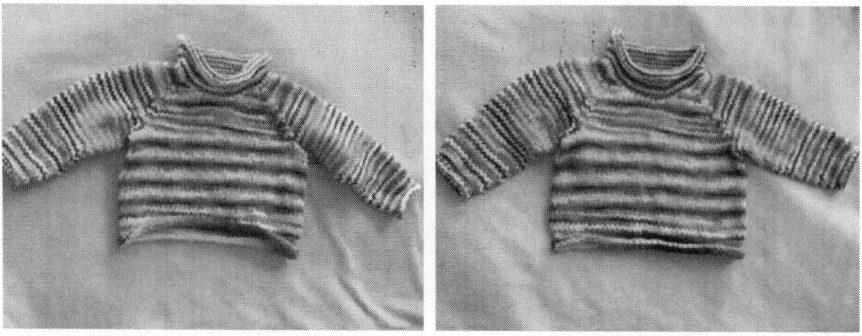

If you have any questions or suggestions, please feel free to contact me!

KNIT A SWEATER

If you want to learn how to knit a sweater you've come to the right place. The goal of this website is to teach you how to make a sweater using the simple, visual, Sweater 101 methodwith a step-by-step video workshop. The video isn't shot yet. I'm still uploading old instructional programs of mine so I hope you're patient. I think it will be worth the wait.

Together we will knit a child's size 1 sweater using any worsted weight yarn you have in your stash or would like to try out. There is no charge. You don't even have to buy the book.

You also don't have to knit to my gauge. You will make this sweater using your yarn and favorite needles and you will draw up the pattern using your personal gauge and it will end up the same size and shape as mine!

Here's a sample of the sweater we're going to make. Enough skills to keep you interested and learning, but simple and small enough to complete in a week or less depending on how much knitting time you get.

If you make this sweater step-by-step with me, you will never again look at a sweater pattern in the same way. You will understand it better and be able to adapt it to your gauge and yarn.

How to Knit a Sweater

To knit a sweater is to feel a great sense of accomplishment. A sweater may seem overwhelming, but all you need are some instructions, the necessary materials and practice. Begin with a mini sweater for a teddy bear. This will help you learn the basic procedures without wasting a lot of time and money. Once you master this miniature version, you can adjust the proportions to make different sizes.

Instructions

1. Begin the Back and Front of Your Teddy Bear Sweater

1. Cast on 134 stitches.
2. Begin the first row by purling one stitch.

- Sponsored Links
- Sewing Machine

3. Create a slip stitch for the second stitch in row one. Place the yarn in front, insert the right needle into the next stitch from back to front and slip the stitch off. Repeat until the end of the row.
4. Begin the second row by placing yarn in front. Insert your right needle into the stitch of the last row from back to front, then slip the stitch off.
5. Continue the second row with a purl stitch while picking up both loops. Repeat for the rest of the row.
6. Repeat two through five until you have 1 inch of fabric.
7. Bind off three stitches for the armhole.
8. Knit 26 stitches for the front as in rows one and two.
9. Bind off three stitches and finish the back.

2. Continue the Back and Front of Your Teddy Bear Sweater

10 Work on the front of the neck at the edge. For every other row bind off two stitches, then one stitch, and then the rest of the stitches for the shoulder.

11 Shape the neck back in the same way.

12 Continue knitting the back until you have an additional 3 inches of fabric.

13 Bind off the 10 middle stitches.

14 Create the neck back at the edge. On every other row bind off two stitches, then one stitch, then all remaining stitches to create the shoulder.

3. Knit the Sleeves of Your Teddy Bear Sweater

15 Cast on 24 stitches in the round.

16 Create the first row. Knit one stitch, purl one stitch, and repeat for the rest of the row.

17 Increase by one stitch for the next row. Repeat the previous step.

18 Repeat the previous row by increasing by one stitch every third row.

19 Repeat the increase five times until you have 29 stitches.

20 Continue until you reach a fabric length of 2 to 2 1/2 inches, then bind off loosely.

4. Finish Your Teddy Bear Sweater

21 Sew the shoulder seams.

22 Set in the sleeves.

23 Pick up 39 stitches in the round for the neckband. Follow the pattern described above.

24 Bind off loosely when the neckband is 1/2 inch high.

Very few art or craft activities offer more opportunities for creative expression than painting on fabric, especially for wearable art. The textile paints formulated today are available in a wide array of brilliant colors; they're also water-based, nontoxic and perfectly safe for even young children to use under adult supervision. And children love to paint on fabric. If you're stumped for ideas, just bring in your kids (or borrow a neighbor's child!) and set them loose in the studio. Soon you'll have yards of beautiful, handpainted fabric to use in your next art-to-wear project.

CHOOSING & PREPARING FABRICS

You can paint on just about any textile, and experimentation with various materials is always encouraged, but here are some suggestions for selecting fabrics to paint:

For the smoothest application of paint, choose tightly woven materials. For looser applications, wetinto- wet techniques, or painted designs that don't require precise designs, loosely woven fabrics (with warp and weft threads clearly visible) are perfectly acceptable.

Some artists prefer to paint on silk, while others favor quilter's quality muslin or 100-percent cotton with a high thread count. Still others like to paint on velvet, rayon, polyester, and other materials. The choice is really up to you.

Try obtaining samples of different textiles in various weights and thread counts to make a swatch notebook. Mail-order fabric-supply companies will often provide these swatches free or for a nominal fee. Apply paint to the swatches and store them dry in a notebook where you can also note how much of what type of paint was applied to each swatch.

If you're just getting started, try using 100-percent cotton fabric with a high thread count, and expand your horizons from there. And remember that many painting techniques are suitable for dark or black fabrics, as well as white.

To prepare cotton or muslin fabric, you may prewash it in hot water and dry it on the hottest setting to remove any sizing on the fabric and to pre-shrink it- or you may choose not to pre-wash at all. It's entirely up to you.

Silks must be treated differently and it is advisable to consult an expert before pre-washing; discuss this with the storeowner where you purchased your fabric or with mail-order fabric-supply companies, which are often staffed by knowledgeable fabric artists.

Choosing Paints

The sheer number of waterbased paints suitable for fabric painting can be overwhelming to the beginning artist. There are sheer, translucent paints, opaque paints, metallic, interference and pearlescent paints, acrylic paints and paints formulated specifically for textiles. And if you can't find textile paint in the color of your choice, you can always mix a little textile medium in any acrylic to create textile paint.

Acrylics are available in jars, tubes, and bottles; they can be mixed to create new colors or diluted with water or with acrylic or textile mediums. Applied straight to fabric they will dry stiff, so I recommend mixing in a little textile medium first to create a paint that will dry with a soft finish.

Textile paints are made specifically for fabric painting (although they can be applied to other surfaces as well). If you're just getting started, I recommend trying textile paint. Jacquard's Textile, Neopaque, and umiere paints are excellent choices for beginners and professionals alike, as is Dr. Ph. Martin's ReadyTex paint.

For "on the surface" 3-dimensional techniques, applicator-tipped paints such as Tulip, Plaid and Jones Tones are fantastic.

Painting Tools & Equipment

You really don't need lots of expensive tools or equipment for fabric painting-in a pinch you can even finger-paint! But it is helpful to have at hand a few tools for experimentation.

- Paintbrushes: both flat and round-tip, in various sizes
- Foam brushes: inexpensive, for smoothest application
- Sponges: manmade and sea sponges
- Toothbrushes: for spattering on paint
- Plastic buckets: for rinsing out brushes and sponges as you work

- Plastic covering: to protect the work surface
- Apron: to protect your clothing
- Plastic or Styrofoam plates: for disposable painter's palettes
- Spray bottle
- Paper towels
- Brayer
- Kosher salt

Preparing to Paint

Don protective clothing and/or disposable gloves and cover the work surface (and surrounding floor) with plastic. Pre-wash fabrics as needed, and fill two buckets with fresh water. Place all tools and paints nearby, ready to use. Tear the fabric into pieces or lay out yardage on a plastic-covered table.

BIBLIOGRAPHY

- A wale, according to Knitting Technology: a Comprehensive Handbook and Practical Guide, is a predominantly vertical column of needle loops generally produced by the same needles at successive (not necessarily all) knitting cycles. A wale starts as soon as an empty needle starts to knit (Spencer 1989:17).
- Anonymous (2009-01-21). Knitters turn to graffiti artists with 'yarnbombing'. London: The Telegraph. Retrieved 2009-05-25.
- Bansal, Paritosh (2008-01-24). Designer Von Furstenberg sues Target over dress. Reuters.
- Breen, Fiona (2012-11-04). A new movie, shot by a Tasmanian filmmaker on a set made almost entirely out of yarn, was partially inspired by knitted graffiti. ABC. Tasmania.
- Cohan, Jillian. Women find creativity, comfort in knitting circle. Archived 2007-03-02 at the Wayback Machine The Wichita Eagle (November 9, 2006). Retrieved on 2006-12-13
- Costa, Maddy (2010-10-10). The graffiti knitting epidemic. The Guardian. London.
- Craft Yarn Council of America press release, CYCA News: Knitting & Crocheting Are Hot! craftyarncouncil.com Retrieved on 2006-12-13.
- Hollingworth, Shelagh (1983). The Complete Book of Traditional Aran Knitting. St. Martin's Press. ISBN 978-0-312-15635-0.
- Høxbro, Vivian (2004). Shadow Knitting. Loveland, CO: Interweave Press. ISBN 978-1-931499-41-5.
- IT'S OFFICIAL Kint Wits Yarn & Wool Shop. Knitwitspenzance.co.uk. Archived from the original on 2009-07-10. Retrieved 2012-10-14.
- Izard, SS; Jorgensen SS (2003). Knitting into the Mystery: A Guide to the Shawl-Knitting Ministry. Morehouse Publishing. ISBN 978-0-8192-1967-1.

- Janelle Masters (2006). Finger knitting. knitty.com. Retrieved 20 September 2016.
- Knitting Basics. Alamac American Knits LLC. 2004. Archived from the original on 2007-02-27. Retrieved 2006-12-27.
- Knitting Needle Sizes. The Fiber Gypsy. Retrieved 3 Dec 2017.
- Leapman, Melissa (2006). Cables Untangled: An Exploration of Cable Knitting. Potter Craft. ISBN 978-1-4000-9745-6.
- Little Hen Rescue (2008). Hens and their jumpers. Retrieved 9 Feb 2011.
- Manning, Tara Jon (2003). Men in Knits: Sweaters to Knit that He WILL Wear. Loveland, Colorado: Interweave Press. ISBN 1-931499-23-3.
- Manning, Tara Jon (2004). Mindful Knitting: Inviting Contemplative Practice to the Craft. Tuttle Publishing. ISBN 978-0-8048-3543-5.
- Morgan-Oakes, Melissa. Toe-Up 2-at-a-Time Socks. Storey Publishing LLC, 2010. ISBN 1-60342-533-0.
- Murphy, Bernadette (2002). Zen and the Art of Knitting: Exploring the Links Between Knitting, Spirituality, and Creativity. Avon, MA: Adams Media. ISBN 1-58062-654-8.
- Obsessed with life's extremes. Waikato Times. 5 October 2007. Archived from the original on February 23, 2013. Retrieved 20 September 2016.
- Pearl-McPhee, S (2005). At Knit's End: Meditations for Women Who Knit Too Much. North Adams, MA: Storey Publishing. pp. 64, 161, 313. ISBN 978-1-58017-589-0.

INDEX

A

According 136

Across 25

Acrylic 255

Active 67, 74

Adding 151

Advanced 239-240

Alpaca 91

Alternate 15, 30

Ancient 157

Another 149, 155

Applied 97

Appliqué 94-95, 97-99

Arabian 185

Around 150, 201

Associated 57

Attach 26

Automatic 128

Available 70, 91, 93, 137

B

Background 155

Balance 121, 124-125

Bamboo 71

Bangladesh 127-129, 132

Became 43, 57, 59

Because 26, 59

Before 48, 69, 73, 161

Beginning 72, 249

Behind 154

Believe 184

Between 16, 31, 162, 166, 191

Binding 162

Blindness 171

Blocked 48

Blocking 46, 48, 251

Bottom 166-167, 191

Brains 171

Buttonhole 243

Buttons 28

C

Cabled 148, 150

Cables 145-147, 149-151, 155-156, 160, 162

Calculate 119

Called 36, 40, 83, 100, 174, 190

Carriage 41, 241

Carried 185

Casting 186, 190

Center 156

Century 51, 53-56, 60, 62-64, 101, 103, 173-174, 185

Certain 122

Charmeuse 114

Chevreul 102

Chicken 92

Chikan 92-93

Children 171-172

Circular 32, 36, 70-71, 75, 77-79, 130, 138-143, 148, 186, 189, 191, 242

Clothes 95

Coarseness 118

Collection 62, 91, 103
Colors 93, 145, 207, 230-231
Colour 102
Colours 57
Column 223
Common 109
Companies 130
Company 136
Comparable 180
Complex 149
Context 97
Continue 253
Continued 64
Controlled 138
Cotton 23, 93, 105-107, 110, 113, 125, 173
Counter 125
Country 128
Counts 115, 122
Covers 93, 95
Create 32, 139, 186, 253, 255
Creates 186, 189
Creating 143
Crewel 93-94
Crochet 90, 172-176, 179-181
Crocheted 174-175, 180
Crocheting 17
Crossed 184
Crossing 154-155
Crotchet 173
Curvature 21
Curved 84
Cylinders 143

D

Darker 230
Darning 199
Decorated 93
Decorative 106
Decrease 212, 215-216, 218
Decreases 213, 216, 219
Denier 109-111, 113
Density 110-111
Depicts 103
Depression 57
Design 132, 137, 140, 213
Designed 84, 104, 148
Designers 139-140, 142-143
Designs 143
Desired 230
Determine 124
Develop 171
Developed 54
Diameter 36, 67, 74, 111, 118
Different 20, 35, 39, 52, 74, 97, 108, 121-122, 142, 178, 190, 213
Direct 119
Displayed 103
Domestic 41
Double 122, 225, 232
Download 44
Duplicate 207
During 52, 57, 59

E

Earliest 184
Easier 60
Easily 71

Egyptian 184
Either 180
Elasticity 21
Embroidery 83, 92-96, 117
Ending 214
England 61
Entrelac 208
Europe 53, 55, 59, 185
Example 35, 39, 62, 178
Exercise 29

F

Fabric 11-13, 16-21, 31, 34, 41-42, 93, 99, 104, 115-116, 121, 125-126, 131, 138, 141-142, 144, 154, 179-180, 185, 189, 209, 212, 227, 231-232, 254-255
Fabrics v, 11, 13, 18, 30-31, 33, 80-81, 131, 134, 138, 140-141, 143-144, 222-223, 231-232, 254
Factories 56
Factors 38
Faggoting 223, 234
Fair-isle 231
Fashion 57, 129, 235
Fassett 64
Features 97
Feeder 135
Felted 32
Felting 32, 189
Fibers 38-39, 47, 58, 86, 104, 110, 177-178
Fibres 141, 144
Filament 110-111
Filaments 110
Fineness 112, 117-118
Finger viii, 163-167, 169, 171-172, 191
Fingers 163, 165-166, 169, 171, 191
Finished 109
Fisherman 158
Fishermen 158
Floreal 129
Following 99, 122, 150
Formation 104
Formula 119
Fringe 204

G

Garment 20, 36, 104, 139, 142, 197
Garments 12, 21, 57
Garter 34
-gauge 122
Gauges 129
Generally 37, 39, 176, 178
Gently 47
Graffiti 257
Griswold 56
Growing 183
Growth 127-128

H

Handknit 47
Hanging 62
Headband 148, 153
Height 180
Higher 115
History vii
Hoffmann 132
Hyperbolic 181

I

I-cord 26-27
Identified 158

Inches 77, 148, 200, 202-205
Include 39, 148, 178
Increase 215, 247-248
Increases 19, 219
Indicates 109
Indirect 119
Industry 56, 131, 134
Installed 128
Internet 60
Invention 56
Involved 151
Ireland 158
Island 157
Islands 158

J

Jacket 62
Jacquard 102
Jumper 156-157
Jumpers 157-158

K

Kantha 95-96
Keeping 247
Kneeling 98
Knitted 11-13, 16-22, 31-34, 36, 40, 47, 51, 53-55, 59, 61-62, 64, 80-81, 143, 148, 154, 180, 184, 186, 190, 196-197, 209, 212, 222, 225, 227, 231-232, 243-244
Knitter 186, 243
Knitters 33, 72, 197-198, 239
Knitting v, vi, vii, viii, ix, 11-13, 17, 19-23, 28, 31-36, 38, 40-44, 46, 48-49, 51-64, 67-70, 73-78, 80, 121-122, 127-135, 137-148, 154, 157, 159, 163-164, 168, 171-172, 179-180, 183-192, 197-201, 205-213, 222-223, 225, 227-228, 231-232, 234, 236, 238-243, 257-258
Knitwear 57, 81, 127
Kurtas 93

L

Larger 244
Laurent 235
Layers 99
Leather 84
Left-hand 69
Length 39, 67, 71, 74, 113, 117-120, 123-124, 177
Lengths 77
Little 228
Loosely 254

M

Machine 42, 56, 96-97, 99, 108, 122-123, 130-133, 135-139, 142, 242
Machines viii, 41, 56, 121-122, 128-132, 134, 139-143, 241
Margaret 213
Marker 249
Markers 245-246
Market 93, 130, 132, 134
Material 71, 94, 96
Materials 78, 94, 175, 254
Mattress 135
Measure 29, 113, 118
Measured 110
Measuring 111
Metallic 106
Meters 109-111
Method 22, 35, 77, 118, 187-188, 190, 196, 243
Methods 17, 190

Middle 166, 191

Million 129

Mirror 95

Mirrors 95

Models 97

Modern v, vi

Moisture 47

Mommes 113

Mosaic 228, 230-231

Multiple 35

Museum 103

N

Necessary 47

Neckband 250

Needle 11, 21, 24, 26, 32, 35-36, 41, 44-46, 67-71, 73-78, 80, 83-87, 89-90, 92, 100, 121, 123, 133, 141, 144-145, 147-148, 154, 159, 161, 186, 189-190, 192-196, 199, 205-206, 216-218, 227, 236-238, 243-244, 246, 248-249

Needlelace 100

Needles 23, 32, 36, 40-41, 44, 60, 67-68, 70-80, 83-87, 92, 148, 163, 186, 189, 199, 205, 210, 236-237

Needlework 89-90, 98

Neighbor 156

Normally 150

Nottingham 43

Novels 198

Number 107, 109-110, 112-116, 118, 123, 154, 160, 224

Numbered 229

Numbering 118

Numerical 117

O

Occasions 93

Offers 135-136

Online 97

Options 131

-over- 150-151

P

Packet 85

Padded 47

Pailung 131

Painting 255

Paints 255

Parameters 122, 126

Passed 16, 31

Pattern 23, 40, 69, 87, 136, 148, 150-151, 156, 161-162, 202-203, 211, 213, 221, 227, 231

Patterning 228

Patterns viii, 19, 33-34, 52, 57, 64, 149, 155-157, 159, 174, 207, 212, 215, 219, 228, 231, 243

People 59, 185

Person 132

Phulkari 96

Picking 190

Pieces 22, 32, 79, 242

Piecing 107

Pilati 235

Placed 125

Plastic 72, 256

Plymouth 91

Pockets 200

Pointer 125

Polyester 105-106

Popular 64
Popularity 58
Position 127
Possible 15, 30
Posture 98
Previous 14
Printout 30
Probably 185
Process 104, 144, 187
Processing 104
Produce 31, 40-41
Produced 42, 232, 234
Production 138
Products 129
Program 99
Project 45, 71, 87, 207
Projects 76, 79, 206
Provides 93
Publishing 258
Pulling 187
Purchase 97
Purled 227

Q

Quickly 72
Quilters 108
Quilting 83, 98, 105, 108
Quilts 98

R

Rather 58
Really 201
Reference 122
Remove 47
Repeat 202, 214, 230, 253
Represents 119
Require 231
Retrieved 257-258
Reverse 233
Ribbing 34, 48, 223-225
Right-side 150-151

S

Sample 119, 125
Samples 173
Seamless 139, 142
Search 211
Second 230, 253
Sector 127
Secure 22
Segment 127
Selection 122
Separately 20
Sewing 83-84, 86, 90, 96-97, 108, 117
Shapes 95
Shaping 25
Sharps 83
Sheesha 95
Shetland 209, 212
Shirts 93
Should 166, 191
Similar 84, 180
Simple 40, 149, 228
Single 180
Sizing 107
Skeins 37, 176
Skipped 153
Skipping 153
Slants 212

Slipped 17, 31, 216, 225, 227-228, 231
Slippery 72
Sliver 144
Smaller 249
Smooth 72
Social 197
Society 63
Software 137
Something 235
Sometimes 39, 178
Spinning 177
Sponges 255
Spread 185
Square 114
Standard 109, 160, 179, 241
Started 184
Stitch 12, 14-17, 22-24, 26-27, 30-31, 34, 36, 38, 41-46, 54-55, 67, 69, 74, 99, 136, 154-155, 161-162, 179-181, 185, 187-188, 190, 192-196, 199, 203, 207, 216-217, 222-225, 227, 230-231, 233-234, 242-243, 245, 248, 253
-stitch 243
Stitched 99
Stitches 13, 15-22, 25, 27, 30-31, 34-36, 40-41, 48, 65, 67-72, 74-75, 81, 94, 100, 145-147, 149-151, 154, 158-162, 165, 175-176, 179-180, 185-186, 188, 190, 192, 196-197, 199-200, 204-205, 207, 213, 215, 218, 220, 223-228, 230-231, 234, 236-238, 242-243, 245, 250, 253
Stockings 52, 54-55
Straight 70, 79, 84
Strand 110
Stranded 207
Strands 109
Stretch 21, 33
Stripes 207, 224
Suitable 122
Surface 47
Survive 173
Swatch 73, 162
Sweater 22, 128-129, 156, 161, 198, 243-244, 252-253
Sweaters 54, 81, 128-129, 159
Synthetic 39, 178
System 107, 109, 118-120, 140, 143
Systems 71, 143

T

Tambour 173
Tapered 74
Tapestries 100-103
Tapestry 84, 98, 100-104
Technical 138
Technique 53, 59, 163, 189, 207, 225, 228, 236
Techniques 44, 46, 48, 89, 239, 242
Technology v, vi, 127, 130, 134
Template 124
Tension 38
Textile v, 90, 96, 117, 136, 255
Textiles vi, 138
Texture 34
Textures 34
Thicker 37, 39, 71, 178
Thickness 39, 177
Thoroughly 47
Thread 83, 89, 100, 105-110, 113-115, 135, 173, 254
Threading 83

Threads 11, 21, 100, 102, 105-109, 114-115

Through 16-17, 31, 36, 41, 47, 67, 74, 83, 97, 171, 187, 190

Throughout 185

Tighten 167, 192

Together 17, 37, 105, 109, 163-164

Tracking 45

Transfer 141, 143

Transition 56

Tricot 22, 223, 236

Tubular 77

Turning 160

Typically 106

U

Ultra-fine 143

Unknitting 45

Usually 40, 75, 155, 157, 189

V

Variety 39, 100, 178

Various 20, 77

Vertical 231

Virtual 137

Volume 120

W

Walker 228

Warmest 200

Washing 47

Weaving 21, 100, 110

Weight 37, 91, 107-109, 111, 113-114, 119, 125

Whereas 142

William 55

Within 94

Without 145

Wooden 72

Woolen 104

Worked 34, 145, 150-151, 179, 229

Working 45, 151, 165, 216

Worsted 104, 112

Wrapped 25, 44, 109

Wrong-side 151

Y

Yarnbank 137